50th IMO – 50 Years of International Mathematical Olympiads

Hans-Dietrich Gronau • Hanns-Heinrich Langmann

Dierk Schleicher

Editors

50th IMO – 50 Years
of International Mathematical
Olympiads

 Springer

Editors

Hans-Dietrich Gronau
Universität Rostock
Institut für Mathematik
Ulmenstr. 69
18051 Rostock
Germany
gronau@uni-rostock.de

Hanns-Heinrich Langmann
Bildung & Begabung gemeinnützige GmbH
Bundeswettbewerb Mathematik
Kortrijker Str. 1
53177 Bonn
Germany
langmann@bundeswettbewerb-mathematik.de

Dierk Schleicher
Jacobs University
Research I
Campus Ring 1
28759 Bremen
Germany
dierk@jacobs-university.de

ISBN 978-3-642-14564-3   ISBN 978-3-642-14565-0 (eBook)
DOI 10.1007/978-3-642-14565-0
Springer Heidelberg Dordrecht London New York

Mathematics Subject Classification (2010): 00A09

Cover design: WMXDesign GmbH, Heidelberg

Printed on acid-free paper

Springer is part of Springer Science+Business Media (www.springer.com)

Preface

From 10 to 22 July 2009 Germany hosted the 50th International Mathematical Olympiad (IMO). This jubilee anniversary event took place on the campus of Jacobs University Bremen. For the first time the number of participating countries exceeded 100, with 104 countries from all continents sending a delegation consisting of six students, a leader and a deputy leader.

Celebrating the 50th anniversary of the IMO provides an ideal opportunity to look back over the past five decades and to review its development from modest beginnings to become a worldwide event. This book is aimed at students in the IMO age group and seeks to demonstrate that mathematics is an active and lively subject, thousands of years old and yet young and fresh as ever. We have invited several IMO pioneers to tell us about the development of IMO since it first began: they reveal that initially nobody thought it would develop into an annual event, let alone grow as big as it is today.

We include three accounts from IMO pioneers: Mircea Becheanu from Romania, speaking on behalf of the country that created the IMO and hosted the initial Olympiad event; István Reiman from Hungary who personally experienced the early years; and Wolfgang Engel from (then East) Germany who witnessed the impact the Olympiads have had from the outset on mathematical activities for students, particularly in Germany. These reports give a vivid and personal description of what gradually became an international success story.

Also included is data about the 50 annual Olympiads which illustrates the development of the competition from just seven countries in 1959 to 104 countries, with 565 participants, in 2009. We list the most successful contestants, the results of the 50 Olympiads and the names of all 112 countries that have ever taken part.

In mathematics, talent is often developed and demonstrated at an early age. Thus it is impressive to see that many of the world's leading research mathematicians were among the most successful IMO participants in their youth. To illustrate this we include a list showing winners of prestigious research awards who participated in the IMO when they were young.

In addition to the mathematics competition an important aspect of the Olympiads is that students from all over the world meet and exchange ideas. It is both a personal highlight of their lives that they treasure for a long time and a place where international friendships develop. Some of the personal or amusing anecdotes of previous IMOs are included in the last chapter of the book.

The outstanding event of the 2009 Olympiad was the celebration of the IMO's 50th anniversary. Six of the world's leading mathematicians, all of them highly successful former IMO participants, were invited as guests of honour to give presentations: Béla Bollobás, Timothy Gowers, László Lovász, Stanislav Smirnov, Terence Tao and Jean-Christophe Yoccoz. Their contributions are also included in this book.

We would like to express our sincere gratitude to all those who supported the organisation of the 50th IMO, making it a memorable and unique event for all participants.

Finally, we wish to thank the Springer Verlag for publishing this book. We are indebted to Clemens Heine for his support and collaboration.

Rostock, Bonn, Bremen   *Hans-Dietrich Gronau*
May 2010   *Hanns-Heinrich Langmann*
                                                                    Dierk Schleicher

Welcome Message

Dear IMO participants,

This year's IMO has a special significance: we celebrate the 50th anniversary of the IMO. In the beginnings it started as a one-time event for students from the countries which at the time were said to form the "socialist block" or "Soviet block", to wit: Bulgaria, Czechoslovakia, German Democratic Republic, Hungary, Poland, Romania and the Soviet Union. It is remarkable that of those 7 countries, 3 do not exist any more.

But the IMO, which, encouraged by the success of the first competition became a yearly event, flourishes. Some key dates: 1967 — the first West-European countries (United Kingdom, France, Italy, Sweden) take part in the 9th IMO in Cetinje, Yugoslavia. (Finland's one-time participation in 1965 had no immediate continuation.) The Netherlands followed suit in 1969, Austria in 1970. Two non-European socialist countries participated in the 60's and early 70's: Mongolia since 1964 and Cuba since 1971. The 16th IMO in 1974 in Erfurt (East Germany) saw a true breakthrough: the teams of the USA and of Vietnam participated for the first time.

Lienz (Austria) was the venue of the first "Western" IMO in 1976, Washington, DC of the first non-European IMO in 1981. Australia organized the 1988 IMO in Canberra, the first Asian IMO took place in 1990 in Beijing, China. Argentina was the venue of the 1997 IMO; we are looking forward to an IMO on African soil in the not-too-far-away future.

My sincere hope is that more and more countries will join the IMO community; old participating countries will see more and more benefit from the IMO for their own mathematical education system and mathematical talent search; and the IMO will continue to be a success story in the next 50 years.

Bremen, July 2009   *József Pelikán*
                                      Chairman of the IMO Advisory Board

József Pelikán

Hungary

József Pelikán took part in 4 IMOs in his student days: In 1963 he won a silver medal, and in 1964-65-66 three gold medals. He was also awarded twice with a special prize for an outstanding solution of a problem.

Since 1971 he has been teaching at Eőtvős Loránd University, Budapest, Hungary. He participated in the organization of the 1970 IMO held in Keszthely, Hungary, and was one of the chief organizers of the 1982 IMO in Budapest, Hungary.

He has been the leader of the Hungarian IMO team since 1988. He was an elected member of the IMO Advisory Board between 1992 and 2002, and has been the Chairman of the IMO Advisory Board since 2002.

Contents

Part I 50th International Mathematical Olympiad – Germany 2009

1   **Committees** 5

    1.1   IMO Advisory Board 5
    1.2   Steering Committee 5
    1.3   Organizers 6
    1.4   Problem Selection Committee 6
    1.5   Jury 7
    1.6   Coordinators 8
    1.7   Invigilators 9
    1.8   Guides 10
    1.9   50th IMO Anniversary Committee 12
    1.10 Volunteers 12

2   **Regulations** 13

3   **Carl Friedrich Gauss, Curvature, and the Cover Art of the 50th IMO** 21

4   **Programme** 27

    4.1   IMO Programme Overview 27
    4.2   Programme Details 28
    4.3   Daily News 32

5   **Participants** 47

    5.1   National Teams 48
    5.2   Observer Countries 100

6   Mathematical Competition 101

    6.1   Problems 103
    6.2   Solutions 113
          6.2.1   Problem 1 – Number Theory – Australia 113
          6.2.2   Problem 2 – Geometry – Russia 116
          6.2.3   Problem 3 – Algebra – United States 118
          6.2.4   Problem 4 – Geometry – Belgium 120
          6.2.5   Problem 5 – Algebra – France 122
          6.2.6   Problem 6 – Combinatorics – Russia 123
    6.3   Awards 126
          6.3.1   Gold Medals - Top Scores 126
          6.3.2   Gold Medals 126
          6.3.3   Silver Medals 127
          6.3.4   Bronze Medals 129
          6.3.5   Honourable Mentions 131
    6.4   Individual Scores 134

7   Mathematical Guests of Honour 155

    7.1   Béla Bollobás 156
    The Lion and the Christian, and Other Pursuit and Evasion Games 157

    7.2   Timothy Gowers 169
    How do IMO Problems Compare with Research Problems? 171

    7.3   László Lovász 185
    Graph Theory Over 45 Years 187

    7.4   Stanislav Smirnov 197
    How do Research Problems Compare with IMO Problems? 199

    7.5   Terence Tao 210
    Structure and Randomness in the Prime Numbers 211

    7.6   Jean-Christophe Yoccoz 217
    Small Divisors: Number Theory in Dynamical Systems 219

Part II History: 50 Years International Mathematical Olympiads

8   Brief Survey 229

9   All IMOs 233

10 All Countries 235

11   **All Results** 239

12   **Memories** 271
     12.1   Mircea Becheanu 271
     12.2   István Reiman 274
     12.3   Wolfgang Engel 277
     12.4   Radu Gologan 282

13   **Hall of Fame** 285

14   **The Golden Microphone** 291
     Rafael Sánchez Lamoneda

A   **IMO Country Codes** 295

Contributors

The following IMO enthusiasts contributed to various texts and tables in this volume:

Hans-Dietrich Gronau
Institute of Mathematics, University of Rostock, D-18051 Rostock, Germany,
e-mail: gronau@uni-rostock.de

Hanns-Heinrich Langmann
Bildung & Begabung gemeinnützige GmbH, Kortrijker Str. 1, 53177 Bonn,
Germany, e-mail: langmann@bundeswettbewerb-mathematik.de

Dierk Schleicher
Jacobs University Bremen, Research I, Postfach 750561, D-28725 Bremen,
Germany, e-mail: d.schleicher@jacobs-university.de

Roger Labahn
Institute of Mathematics, University of Rostock, D-18051 Rostock, Germany,
e-mail: roger.labahn@uni-rostock.de

Matjaž Željko
Faculty of Mathematics and Physics, University of Ljubljana, SI-1000 Ljubljana,
Slovenia, e-mail: matjaz.zeljko@fmf.uni-lj.si

Anke Allner
Jacobs University, Postfach 750561, D-28725 Bremen, Germany
e-mail: a.allner@jacobs-university.de

Book layout: Roger Labahn with LaTeX & TeX

We wish to thank Jane Ferguson (London) for carefully revising selected sections of the English text, as well as Eckard Specht (Magdeburg) for producing all the geometric drawings in Section 6.2.

Part I
50th International Mathematical Olympiad
– Germany 2009

The IMO 2009 was supported by

GEFÖRDERT VOM

The IMO 2009 was organised by

in cooperation with

Main sponsors:

Sponsors:

We gratefully acknowledge additional support from:

BEGO Bremer Goldschlägerei, Bremer Messegesellschaft HGV, Die Sparkasse Bremen AG, GEOtec Zeichen- und Kunststofftechnik GmbH, Hannover Rückversicherung AG, Mathematikum e.V., NORDMILCH AG, Verkehrsverbund Bremen/Niedersachsen GmbH (VBN), Zirgon GmbH

Chapter 1
Committees

1.1 IMO Advisory Board

Chairman

József Pelikán (Hungary)

Secretary

John Webb (South Africa)

Elected Members

Gregor Dolinar (Slovenia)
Myung-Hwan Kim (Republic of Korea)

Patricia Fauring (Argentina)

Co-opted Members

María Gaspar (Spain 2008)
Hans-Dietrich Gronau (Germany 2009)

Tildash Bituova (Kazakhstan 2010)
Wim Berkelmans (Netherlands 2011)

1.2 Steering Committee

Anke Allner (Jacobs University Bremen)
Hans-Dietrich Gronau (Jury Chairman)
Hanns-Heinrich Langmann (Director IMO-Office Bonn)
Dierk Schleicher (Bremen; Anniversary Celebration)
Harald Wagner (Executive Director)

1.3 Organizers

IMO2009 – Office Bonn

Menna Jones   Harald Wagner
Hanns-Heinrich Langmann   Tanja Weck
Eva Vahjen

IMO2009 – Office Bremen

Anke Allner   Cristian Nucu Mesaros
Roselies Cordes   Georg Schröter
Frauke Dammann   Silke Tilgner
Jane Ferguson   Oliver Walenziak
Daniel Hockenberry

IMO2009 – IT

Roger Labahn   Matjaž Željko

1.4 Problem Selection Committee

Konrad Engel   Jürgen Prestin
Karl Fegert   Christian Reiher
Andreas Felgenhauer   Peter Scholze
Hans-Dietrich Gronau   Eckard Specht
Roger Labahn   Robert Strich
Bernd Mulansky   Martin Welk

1.5 Jury

Host Country Representatives

Hans-Dietrich Gronau (Jury Chairman)   Roger Labahn (Secretary)
Jürgen Prestin (Chief Coordinator)

Team Leaders

Fatmir Hoxha (ALB)
Abed-Seddik Bouchoucha (ALG)
Patricia Fauring (ARG)
Nairi Sedrakyan (ARM)
Angelo Di Pasquale (AUS)
Robert Geretschläger (AUT)
Fuad Garayev (AZE)
Mahbub Alam Majumdar (BGD)
Igor Voronovich (BLR)
Bart Windels (BEL)
Assogba Bernardin Kpamegan (BEN)
René Arturo Aguilar Vera (BOL)
Damir Hasić (BIH)
Carlos Yuzo Shine (BRA)
Nikolai Nikolov (BGR)
Lin Sok (KHM)
Dorette Pronk (CAN)
Hernan Burgos (CHI)
Hua-Wei Zhu (CHN)
Maria Elizabeth Losada (COL)
Mario Alberto Marín Sánchez (CRI)
Mea Bombardelli (HRV)
Eduardo Pérez Almarales (CUB)
Andreas Skotinos (CYP)
Jaroslav Švrček (CZE)
Jens-Søren Kjær Andersen (DEN)
Jorge Chamaidan (ECU)
Härmel Nestra (EST)
Matti Lehtinen (FIN)
Claude Deschamps (FRA)
Larry Gogoladze (GEO)
Eric Müller (GER)
Anargyros Fellouris (HEL)
Javier Ronquillo (GTM)

Mario Roberto Canales Villanueva (HND)
Tat Wing Leung (HKG)
József Pelikán (HUN)
Auðun Sæmundsson (ISL)
Chudamani R Pranesachar (IND)
Hery Susanto (IDN)
Arash Rastegar (IRN)
Bernd Kreussler (IRL)
Shay Gueron (ISR)
Roberto Dvornicich (ITA)
Yuji Ito (JPN)
Yerzhan Baisalov (KAZ)
Yong Chol Ham (PRK)
Myung-Hwan Kim (KOR)
Ibrahim Alqattan (KWT)
Buras Boljiev (KGZ)
Agnis Andžāns (LVA)
Julian Kellerhals (LIE)
Artūras Dubickas (LTU)
Charles Leytem (LUX)
Ieng Tak Leong (MAC)
Vesna Manova-Erakovik (MKD)
M. Suhaimi Ramly (MAS)
Isselmou Ould Lebat Ould Farajou (MRT)
Radmila Bulajich (MEX)
Valeriu Baltag (MDA)
Dashdorj Tserendorj (MNG)
Svjetlana Terzić (MNE)
Abdelmoumen Med El Rhezzali (MAR)
Quintijn Puite (NLD)
Michael Albert (NZL)
Samson Olatunji Ale (NGA)

Dávid Kunszenti-Kovács (NOR)

Alla Ditta Choudary (PAK)

Jaime Gutierrez (PAN)

Jose Guillermo von Lucken (PAR)

Emilio Gonzaga Ramírez (PER)

Ian June Luzon Garces (PHI)

Andrzej Komisarski (POL)

Joana Teles (POR)

Luis Caceres (PRI)

Radu Gologan (ROU)

Nazar Agakhanov (RUS)

Carlos Mauricio Canjura Linares (SLV)

Đorđe Krtinić (SRB)

Yan Loi Wong (SGP)

Vojtech Bálint (SVK)

Gregor Dolinar (SVN)

David Hatton (SAF)

María Gaspar (ESP)

Chanakya Janak Wijeratne (LKA)

Paul Vaderlind (SWE)

Anna Devic (SUI)

Abdullatif Hanano (SYR)

Shu-Chung Liu (TWN)

Umed Karimov (TJK)

Paisan Nakmahachalasint (THA)

Indra Haraksingh (TTO)

Seifeddine Snoussi (TUN)

Okan Tekman (TUR)

Erol Aslan (TKM)

Bogdan Rublov (UKR)

Mohamed Salim Almakhooli (UAE)

Geoff Smith (UNK)

Zuming Feng (USA)

Leonardo Lois (URY)

Shuhrat Ismailov (UZB)

Rafael Sánchez Lamoneda (VEN)

Hà Huy Khoái (VNM)

Thomas Masiwa (ZWE)

1.6 Coordinators

Chief Coordinator

Jürgen Prestin

Problem Captains

Konrad Engel

Karl Fegert

Andreas Felgenhauer

Bernd Mulansky

Robert Strich

Martin Welk

Coordinators

Julian Arndts

Richard Bamler

Christian Bey

Wolfgang Burmeister

Nico Düvelmeyer

Ulrich Derenthal

Michael Dreher

Thomas Fischer

Jan Fricke

Marlen Fritzsche

Frank Göring

Hans-Gert Gräbe

Cornelius Greither
Darij Grinberg
Natalia Grinberg
Paul Jonas Hamacher
Heiko Harborth
Martin Härterich
Sven Hartmann
Klaus Henning
Albrecht Heß
Daniel Herden
Armin Holschbach
Thomas Jäger
Jörg Jahnel
Thorsten Kleinjung
Albrecht Kliem
Norbert Koksch
Philipp Lampe
Martin Langer
Bodo Laß
Uwe Leck
Volkmar Liebscher
Bing Liu
Hendrik Lönngren
Wolfgang Moldenhauer
Stephan Neupert
Monika Noack
Matthias Ohst

Martin Olbermann
Florian Pfender
Gerlind Plonka-Hoch
Daniel Potts
Wolfgang Radenbach
Christian Reiher
Frank Ristau
Karsten Roeseler
Jürgen Roßmann
Michael Rüsing
Jan-Christoph Schlage-Puchta
Peter Scholze
Georg Schönherr
Reinhard Schuster
Stefan Schwarz
Eckard Specht
Colin Stahlke
Gabriele Steidl
Jakob Stix
Kristin Stroth
Matthias-Torsten Tok
Eberhard Triesch
Peter Wagner
Matthias Warkentin
Elias Wegert
Rolf Zimmermann

1.7 Invigilators

Chief Invigilator

Dierk Schleicher

Invigilators

Stephan Baier
Diego Bevers-Schleicher
Jan-Hendrik Bruns
Jan Cannizzo
Daniel Claes
Zymantas Darbenas

Dzmitry Dudko
Jan-Oliver Fröhlich
Gwyneth Gardiner
Tony Gardiner
Stefan Halverscheid
Stanislav Harizanov

Elena Kalinka
Gudrun Kämper
Birgit Krah
Felix Krahmer
Tim Kröger
Klaus Lies
Petra Lies
Marlies Lohse
Keivan Mallahi-Karai
Ingrid Meister-Tolksdorf
Reiner Meister
Yauhen "Zhenya" Mikulich
Sönke Mittwollen
Rüdiger Olbrich

Marcel Oliver
Peter Oswald
Tobias Preußer
Peter Rashkov
Frederik van Schagen
Friedemann Schleicher
Nikita Selinger
Anton Suharev
Henning Thielemann
Michael Thon
Vladlen Timorin
Wiebke Trumann
Sergiy Vasylkevich
Christopher de Vries

1.8 Guides

Chief Guides

Anke Allner
Wilfried Kurth

Stefanie Schiemann

Senior Guides

Jan-Hendrik Bruns
Jan Cannizzo
Zymantas Darbenas
Jan-Oliver Fröhlich
Stanislav Harizanov
Gudrun Kämper
Felix Krahmer

Klaus Lies
Keivan Mallahi-Karai
Rüdiger Olbrich
Peter Rashkov
Nikita Selinger
Anton Suharev
Michael Thon

Team Guides

Shaik Ahmed (PAK)
Meru Alagalingam (LKA)
Marie van Amelsvoort (NLD)
Apostol Apostolov (ARM)
Manuel Bärenz (HND, SLV)
David Bauer (BIH)
Nils Becker (AUT)
Todor Bilarev (BGR)
Lukas Brantner (CAN)
Simon Buchholz (ECU)
Olexiy Chudnovskyy (UKR)
Hauke Conradi (FIN)
Lucie Costard (ISL)
Martin Dieblich (EST)
Clemens Dubslaff (POR)
Jeremias Epperlein (ISR)
Friedrich Feuerstein (LUX)
Hristina Fidanoska (MKD)
Christina Flörsch (TUN)
Tobias Fritz (PRI)
Severin Gierlich (NOR)
Roman Glebov (TJK)
Ingrid von Glehn (SAF)
Tamara von Glehn (ZWE)
Giorgi Gogishvili (GEO)
Matthias Görner (USA)
Ilja Göthel (TKM)
Yue Guan (MAC)
Jan Hackfeld (ALB)
Annika Heckel (TTO)
Andreas Hicketier (CYP)
Svenja Hüning (MRT)
Alin Iacob (LTU)
Florin Ionita (FRA)
Nurazem Kaldybaev (KGZ)
Deborah Kant (URY)
Pauline Koch (COL)
Nozim Komilov (UZB)
Miriam Kümmel (MAS)
Innocent Kwizera (MAR)
Hanna Lagger (ALG)
Thai Le Tran (VNM)
Shu Li (SGP)

Leo Margolis (MNE)
Sarah Marzi (ITA)
Stefan Mehner (HUN)
Pavel Metelitsyn (LVA)
Tobias Mettenbrink (HEL)
Michael Meyer (NGA)
Sergiu Mosanu (TUR)
Tarlan Nazarov (AZE)
Andreas Neuzner (CZE)
Philipp Niemann (NZL)
Sebastian Nill (SWE)
Manuel Nutz (KWT)
Dheeraj Pant (IND)
Ariane Papke (LIE)
Fabian Parsch (HRV)
Artiom Patrinica (RUS)
Vitalie Patrinica (BLR)
Peter Patzt (AUS)
Marina Perich Krsnik (ESP)
Vyacheslav Polonski (POL)
Laura de la Purificación Agudo (PER)
Gunnar Quassowsky (BGD)
Maxim Rauwald (ARG)
Judit Recknagel (GER)
Viktoria Ronge (CHI, CUB)
Eugenia Rosu (UNK)
Rasmus Rothe (DEN)
Nithi Rungtanapirom (THA)
Alexander Sacharow (SRB)
Markus Schepke (PHI)
Tanja Schindler (PAR)
Simon Schmitt (IRL)
Tobias Schoel (SVK)
Sebastian Schwab (BEN)
Marvin Secker (MEX)
Shafie Shokrani (IRN)
Matvey Soloviev (JPN)
Christoph Sommer (BOL)
Eugen Sorbalo (MDA)
Hinnerk Stach (KOR)
Igor Stassiy (KAZ)
Andreas Steenpaß (KHM)
Julia Steinberg (MNG)

Konrad Steiner (SYR)
Svenja Strecker (UAE)
Chenshuai Sui (CHN)
Xiao Sun (TWN)
Stefan Toman (PRK)
Timm Treskatis (IDN)
Antony Trinh (BEL)

Uchenna Udeh (GTM, VEN)
Bogdan Vioreanu (ROU)
Jue Xiang Wang (HKG)
David Willimzig (SVN)
Barbara Wodarz (SUI)
Anne Zander (BRA)
Alraune Zech (CRI, PAN)

Jury Guides

Jessica Fintzen
Amos Schikowsky
Julia Singer

Pavel Zorin-Kranich
Philipp Weiß

1.9 50th IMO Anniversary Committee

Anke Allner
Hans-Dietrich Gronau
Martin Grötschel

Hanns-Heinrich Langmann
Dierk Schleicher
Günter M. Ziegler

1.10 Volunteers

Astrid Bayor
Diego Bevers-Schleicher
Dylan Bevers
Florian Brandt
Dzmitry Dudko
Karst Koymans
Daniel Kurzyk
Yauhen "Zhenya" Mikulich
Joscha Nause
Marcel Oliver
Peter Oswald

Uwe Pagel
Nils Przigoda
Thomas Rieger
Nikita Selinger
Michael Stoll
Marlon Tilgner
Vladlen Timorin
Christopher de Vries
Raymond O. Wells, Jr.
Stefan Wieding
Rob Wieleman

Chapter 2
Regulations

1. General

   1.1. The 50th International Mathematical Olympiad (the "IMO 2009") will be held in Germany from July 10 to July 22, 2009.

   1.2. These regulations (the "regulations") govern the running of the IMO 2009.

   1.3. The Association Bildung und Begabung e.V. ("Bildung und Begabung") has an overall responsibility for the organization of the IMO 2009.

   1.4. The aims of the IMO 2009 are:
   - to discover, encourage and challenge mathematically gifted young people in all countries;
   - to foster friendly international relationships among mathematicians of all countries;
   - to create an opportunity for the exchange of information on school syllabuses and practices throughout the world;
   - to promote mathematics generally.

2. Participation

   2.1. Participation in the IMO 2009 is by invitation only. Each invited country is entitled to send a team con-sisting of up to six contestants (the "Contestants"), a Leader and a Deputy Leader to be known collectively as the participants (the "Participants").

   2.2. Contestants must not have formally enrolled at a university or any other equivalent post-secondary institution, and they must have been born on or after July 17, 1989.

   2.3. Observers including family members (the "Observers") may apply to accompany the Participants, the terms under which they can do so are detailed in sub-clause 3.5 below.

   2.4. The official programme (the "Official Programme") as referred to below is the programme and outline itinerary for the IMO 2009 and associated events. Bildung und Begabung reserves the right to amend or revise the Official Programme in whole or part. If there is by any chance a significant change, Participants and Observers of the invited countries will be notified in advance.

- The Official Programme for Leaders begins on July 10, 2009 and ends on July 22, 2009.
- The Official Programme for Deputy Leaders and the Contestants begins on July 13, 2009 and ends on July 22, 2009.

The Official Programme will contain, among other things, details of accommodation arrangements (including food) for Participants and Observers and the venues for various official events associated with the IMO 2009.

2.5. Each invited country wishing to participate in the IMO 2009 must confirm participation online using username and password provided by Bildung und Begabung no later than February 27, 2009. This will also confirm that the Leader agrees to abide by the Regulations for the IMO 2009.

- Registration of Leaders, Deputy Leaders and Observers must be completed online no later than April 30, 2009.
- Registration of contestants must be completed online no later than June 2, 2009.
- Registration of arrival/departure dates of participants must be completed online no later than June 15, 2009.

2.6. Leaders and Deputy Leaders are responsible for the conduct of the Contestants, and for the avoidance of doubt the Leaders and Deputy Leaders are acting in loco parentis for their Contestants except where Bildung und Begabung has been notified in writing that an Observer has been nominated to act in loco parentis.

2.7. Leaders and Deputy Leaders must ensure that their Contestants know and fully understand clause 5 of these Regulations. They must also make it clear that any Contestant who violates any of these Regulations may be liable to disqualification from the IMO 2009. In order to avoid any trouble or accident, Leaders and Deputy Leaders must also inform fully their Contestants of the contents of "Important Contest Information for Contestants".

3. Responsibility for Accommodation and Expenses

3.1. Bildung und Begabung will provide accommodation in single rooms as a rule for Leaders, Deputy Leaders and Observers and in shared rooms for Contestants. Furthermore Bildung und Begabung will provide meals, transport between Bremen Airport or Bremen Main Railway Station (Bremen Hauptbahnhof) and the accommodation site and other necessary transport between the accommodation site and other venues on the Official Programme for all the Participants and Observers. In order to maintain the integrity of the IMO 2009, the detailed Official Programme will not be disclosed until arrival.

3.2. Other than in respect of the provision of accommodation, meals and transport during the Official Programme as detailed in sub-clause 3.1, Bildung und

Begabung shall not be liable under any circumstances for any costs or expenses whatsoever or howsoever incurred by any Participant or Observer in connection with the IMO 2009. In particular, Bildung und Begabung shall not be liable for any expenses derived from:

- Spending extra days in Germany outside the relevant dates specified in sub-clause 2.4 above;
- Transports to and from Germany incurred by Participants or Observers;
- Transports within Germany prior to arrival at Bremen Airport or Bremen Railway Station or following departure incurred by Participants or Observers.

3.3. All Participants and Observers are responsible for obtaining full accident, health and travel insurance. It is the Leader's responsibility to confirm online using username and password provided by Bildung und Begabung that this condition has been met for all members of his or her team.

3.4. Bildung und Begabung expects to offer opportunities to Participants and Observers for excursions and/or cultural trips but will be under no obligation to do so.

3.5. In reference to sub-clause 2.3 above the deadline for receipt of an application to accompany the Participants as Observer is April 30, 2009. Since extra accommodation is limited, no guarantee is given that such applications will be successful. For those applications that are notified as successful, full payment must be made in cleared funds of the following charges by June 1, 2009. Any application received without the full payment of the charges will be rejected. No refund will be given.

- Observer A accompanying the Leader and residing at or near the Leaders' site: € 2200 (two thousand two hundred Euros) for a single room, € 2000 (two thousand Euros) per person for a double room (for 12 nights).
- Observer B accompanying the Deputy Leader and residing at or near the Deputy Leader' sites: € 1200 (one thousand two hundred Euros) for a single room, € 1000 (one thousand Euros) per person for a double room (for 8 nights).
- Observer C accompanying the Contestants and residing at their site: € 1200 (one thousand two hundred Euros) for a single room, € 1000 (one thousand Euros) per person for a double room (for 8 nights).

Consideration will be given to applications from Observers wishing to attend and observe the IMO 2009 for only part of the period of the Official Programme and in such cases the charges, accommodation and all other relevant arrangements have to be negotiated with Bildung und Begabung.

4. Proposals for Problems

4.1. Each participating country other than the host country is expected to submit up to six proposed problems, with solutions, to be received by the Problem Selection Committee no later than March 31, 2009. Only the Leader may submit the proposals. They should be sent by mail (not by e-mail, for security reasons).

4.2. The proposals should, as far as possible, cover various fields of pre-university mathematics and be of varying degrees of difficulty. They should be new and may not have been suggested for or used in any other Mathematics competition.

4.3. The proposals must only be written in English, French, German, Russian or Spanish. The proposals and solutions should be accompanied by their English versions.

5. Contest Regulations

5.1. The contest element of the IMO 2009 (the "Contest") will take place in Bremen on July 15 and 16, 2009, under the direction of the Chief Invigilator appointed by Bildung und Begabung. On each day of the Contest the examination will start in the morning and last for 4 and a half hours. Each of the two examination papers will consist of three problems.

5.2. Each Contestant may receive the problems in one or two languages, previously requested on the regis-tration form, provided that the Jury (as defined in 6.1) has approved the relevant translation.

5.3. Each Contestant must work independently and submit solutions in his/her own language. The solutions must be written on answer forms provided by Bildung und Begabung. Contestants must write on only one side of each answer form.

5.4. The only instruments permitted in the Contest will be writing and drawing instruments, such as rulers and compasses. In particular, books, papers, tables, calculators, protractors, computers and communication devices will not be allowed into the examination room.

5.5. The Jury, Observers and any others who have sight of the problems and solutions before the examinations shall do their utmost to ensure that no Contestant has information, direct or indirect, about any proposed problem. They must also ensure that all Contest problems and solutions are kept strictly confidential until after the entire Contest has finished. They are barred, from the moment of their arrival in Bremen until the time of conclusion of the second examination, from having any external communication with Contestants, Deputy Leaders and accompanying Observers. In case of an emergency, Bildung und Begabung will provide proper assistance. Similarly, Contestants, Deputies and their Observers are barred from contacting Leaders and Observers accompanying the Leaders, during the same period of time. Information about arrivals, delays about arrivals and similar messages are to be directed exclusively to the published IMO office and may be forwarded by the office to the leaders upon request.

5.6. The total number of prizes will not exceed half the total number of Contestants. The numbers of first, second and third prizes will be approximately in the ratio 1:2:3.

5.7. Special prizes may be awarded for solutions considered outstanding by the Jury. Proposals for such special prizes will be put forward by the Chief Coordinator.

5.8. The prizes will be awarded at the Closing Ceremony. Each Contestant who has not received a first, second or third prize will receive a Certificate of Honourable Mention if he/she has received seven points for the solution of at least one problem.

5.9. Each Contestant will receive a Certificate of Participation.

6. Jury Regulations

6.1. The "Jury" will consist of all Leaders, together with a Chairman. A Leader may be replaced by his/her Deputy Leader in an emergency (subject to the prior approval and consent of the Chairman of the Jury). Members of the IMO Advisory Board who are not already members of the Jury, members of the Problem Selection Committee and the Chief Coordinator (as defined in clause 7 below) may also attend meetings of the Jury as observers. Observers may attend meetings of the Jury only with the permission of the Chairman of the Jury, but will not be entitled to speak or vote. However, they may exceptionally speak at the explicit request of the Chairman of the Jury. Deputy Leaders may attend, as observers, meetings of the Jury held after the Contest.

6.2. Only Leaders may vote in the decisions of the Jury and each Leader will have one vote. A motion shall be carried by a simple majority of those voting. In the event of a tie, the Chairman will have a casting vote.

6.3. The Jury may appoint sub-committees to consider specific matters.

6.4. The meetings of the Jury will be conducted principally in English. The Chairman should request a trans-lation into some of the official languages (French, German, Russian and Spanish) as needed.

6.5. Before the Contest the Jury will
- verify that all Contestants comply with the prescribed conditions for participation;
- select the Contest problems from amongst the submitted proposals on the basis of a preliminary selection made by the Problem Selection Committee appointed by Bildung und Begabung;
- prepare and approve the official versions of the Contest problems in English, French, German, Russian and Spanish;
- approve the translations of the Contest problems into all required languages.

6.6. On each day of the Contest, the Jury will consider written questions raised by Contestants during the first half-hour of the Contest and decide on replies.

6.7. After the Contest, the Jury will
- receive and approve a report made by the Chief Invigilator on the conduct of the examinations;

- receive a report from the Chief Coordinator on any unresolved disputes which may have arisen during coordination and determine the appropriate scores;
- approve the scores of all Contestants;
- decide winners of first, second and third prizes;
- consider and make decisions on all proposals to award special prizes;
- consider matters raised about future International Mathematical Olympiads.

6.8. Any allegation or suspicion of a violation of the Regulations generally shall be reported to the Chairman of the Jury. If he considers there is a prima facie case, he will form a committee to investigate further. The committee will report its findings to the Jury. The Jury will decide whether a violation has occurred and, if it decides that one has, then it will decide what sanction, if any, to apply. Possible sanctions include the disqualification of an individual Contestant or an entire team from the competition. The decision of the Jury will be final.

7. Coordination

7.1. Each problem will be allocated a score out of a maximum of seven points.

7.2. Prior to coordination, Contestants' solutions will be assessed by Leaders and Deputy Leaders in accor-dance with the marking schemes prepared under the direction of the Chief Coordinator appointed by Bildung und Begabung and discussed and agreed upon at the meeting of the Jury.

7.3. Each coordination session will involve two Coordinators provided by Bildung und Begabung and repre-sentatives of the relevant country. Two representatives, normally the Leader and Deputy Leader, are permitted to participate actively in any one session. With the approval of the Chief Coordinator, one further representative may be present to observe the coordination process but cannot take any active part in it.

7.4. The Leader and the designated Coordinators should agree on the scores for each Contestant. These scores will be recorded on official forms and signed by the Leader and the Coordinators. If the Leader and the Coordinators fail to agree on a score for a Contestant, the matter will first be referred to the head Coordinator (Problem Captain) for that problem. If there is still no agreement, the matter will be referred to the Chief Coordinator. If the Leader and Chief Coordinator then fail to agree on a score, the Chief Coordinator will report the matter to the Jury with a recommendation as to what the score should be. The Jury will then determine the score.

7.5. If, during a coordination session, the designated Coordinators consider that an irregularity may have occurred, they will immediately refer the matter to the Chief Coordinator. Unless he is satisfied that there is no case to answer, he will report the situation to the Chairman of the Jury.

7.6. For each problem, solutions by German Contestants will be coordinated by the Leader and Deputy Leader of the country which submitted the problem, with the assistance of the Problem Captain for that problem if required.

8. Entire Agreement and Understanding

8.1. The Leaders, Deputy Leaders and Contestants acknowledge that these Regulations and the documents referred to herein, constitute their entire agreement
and understanding of the parties and supersede any previous discussions or
representations made by or on the behalf of Bildung und Begabung in respect
of the IMO 2009.

9. Force Majeure

9.1. In these Regulations, "force majeure" shall mean any cause preventing Bildung und Begabung from performing any or all of its obligations which arises
from or is attributable to acts, events, omissions or accidents beyond the rea-
sonable control of the party so prevented including without limitation strikes,
lock-outs or other industrial disputes (whether involving the workforce of the
party so prevented or of any other party), act of God, war, riot, civil com-
motion, malicious damage, compliance with any law or governmental order,
rule, regulation or direction, accident, breakdown of plant or machinery, earth-
quake, typhoon, fire, flood, storm, or default of suppliers or sub-contractors.

9.2. If Bildung und Begabung is prevented or delayed in the performance of any of
its obligations to Leaders, Deputy Leaders and Contestants under these Regulations by force majeure it will have no liability in respect of the performance
of such of its obligations as are prevented by the force majeure events during the continuation of such events, and for such time after they cease as is
necessary for Bildung und Begabung to recommence its affected operations
in order for it to perform its obligations.

Chapter 3
Carl Friedrich Gauss, Curvature, and the Cover Art of the 50th IMO

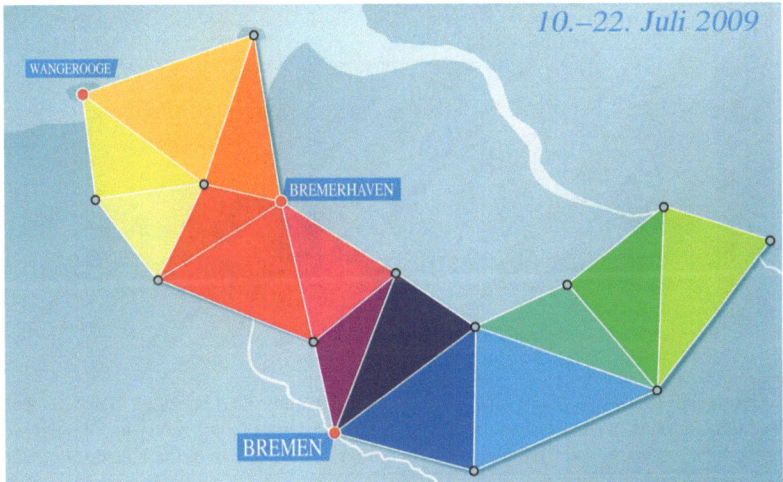

Detail from the IMO 2009 poster showing an artist's rendering of Gauss' map.

The design of the poster of the 50th IMO is an artist's version of a map produced by the German mathematician, physicist, and astronomer Carl Friedrich Gauss in 1821. Gauss' interests relate intrinsic properties of geometry with very practical aspects of numerical calculations and properties of land surveying.

Among mathematicians, Gauss is best known for his fundamental contributions to many areas of mathematics. But he was also a leading physicist (working especially on magnetism) and astronomer (he became particularly famous for his determination of the orbit of the dwarf planet Ceres).

In 1820, Gauss took over the task to survey the territory of the Kingdom of Hannover (much of the current German state of Niedersachsen / Lower Saxony). He was inspired to do this project by his former student Heinrich Christian Schumacher who

The German 10-Mark-bill, in use 1991–2002. The bill features Carl Friedrich Gauss, as well as a graph of the Gaussian normal distribution, several building from Göttingen (where Gauss worked as a professor), a sextant, and (in the lower right corner) part of a map of the Kingdom of Hanover that Gauss produced as a surveyor.

was in charge of surveying the Kingdom of Denmark and suggested that Gauss undertake a similar task in the adjacent country to the South.

Gauss did the surveying measurements by himself, realizing the importance of utmost precision for this work. Sometimes, it is speculated that Gauss made these large-scale measurements partially in order to determine whether our three-dimensional world was Euclidean or not — only in Euclidean geometry is the sum of the three angles in a triangle equal to π (we measure angles in radians: π corresponds to 180°): in spherical geometry, the sum is larger, and in hyperbolic geometry, it is smaller. The deviation of this sum from π is proportional to the area of the triangle (see below). However, Gauss knew very well that any deviation of the sum from π would be far smaller than his experimental error: all his

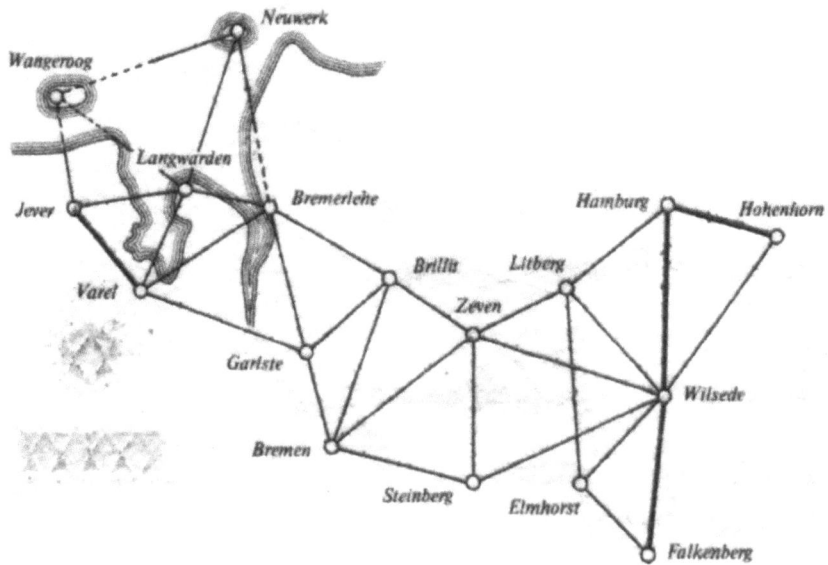

Magnification of Gauss' map of the Kingdom of Hanover from the 10-Mark-bill. This map includes *Bremen*, *Bremerlehe* (now called Bremerhaven) and *Wangeroog* (now called Wangerooge), the three major locations of the 50th International Mathematical Olympiad.

measurements would be compatible with the assumption that our three-dimensional world was Euclidean.

However, the two-dimensional surface of the Earth (embedded in three-space) is *not* Euclidean: it is (approximately) spherical and thus has curvature; and the curvature of this surface is large enough so that it *can* be measured. This is reflected in his experimental data in the fact that for any interior vertex of his measurement grid, the sum of interior angles of adjacent triangles is strictly less than 2π (or 360°). Due to its rotation, the shape of the Earth is not perfectly spherical: the radius is greater near the equator and smaller near the poles. Gauss was aware of these facts and used them in order to improve the accuracy of his measurements — together with the tools of error analysis that he had developed earlier and that he had also taken advantage of in his determination of the orbit of Ceres.

Some of Gauss' most important work is in differential geometry, where he introduced the concept of curvature. Gauss himself revealed that among his primary inspirations for these theoretical geometric investigations were astronomy and (theoretical) surveying. For an oriented smooth surface S in $\mathbb{R}^3$, the scalar curvature that is now known as the Gaussian curvature is defined as follows: for each point $x \in S$, choose a unit vector $n(x)$ perpendicular to S at x. This defines $n(x)$ uniquely up to sign, and orientability of S means that one can define $n(x)$ continuously on all of S. Now $n(x)$ is a point on the unit sphere $S^2 \subset \mathbb{R}^3$; this defines a smooth map $G: S \to S^2, x \mapsto n(x)$. This map is known as the *Gauss map*. The *Gaussian curvature* of S at x, denoted $\kappa(x)$, is the ratio by which the Gauss map expands (signed) areas. On any flat surface, the Gauss map is constant, so the curvature is zero. On a

ZUR NETZAUSGLEICHUNG.

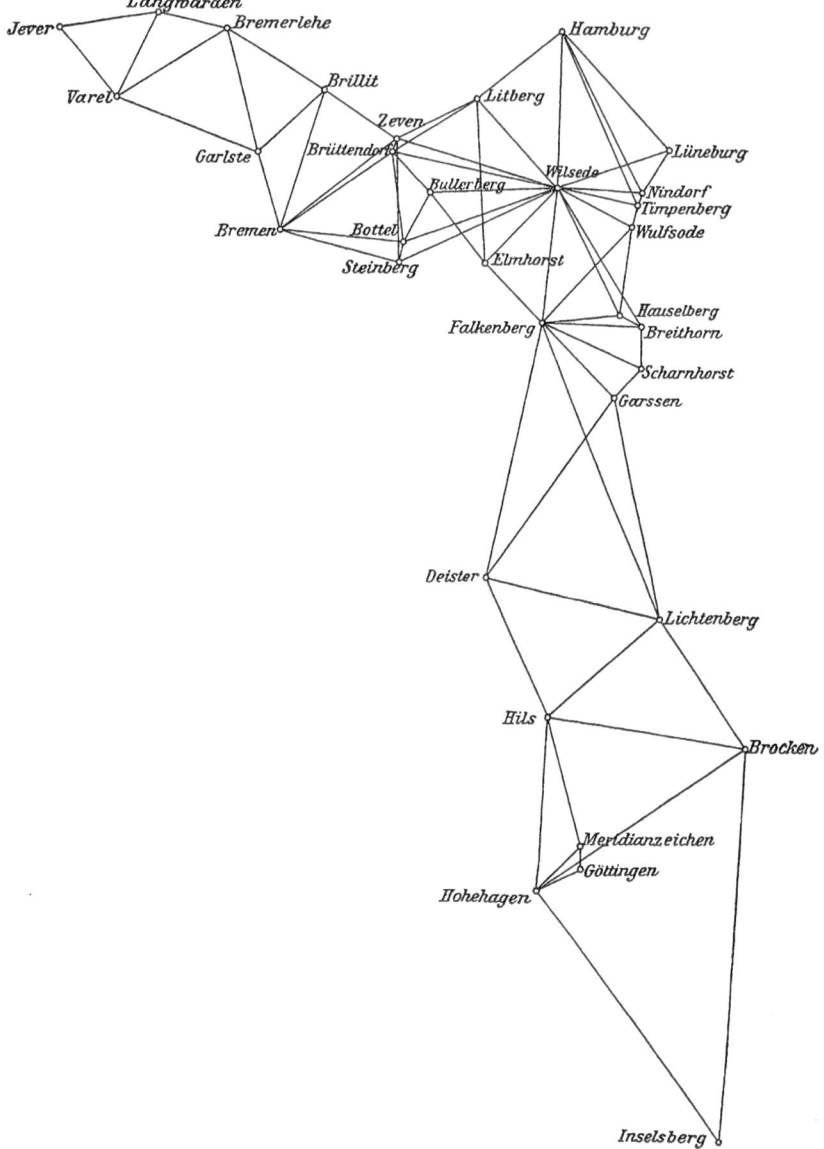

Gauss' original maps from his collected works, extending throughout the kingdom of Hanover. (from: Carl Friedrich Gauß, Werke, Neunter Band. Herausgegeben von der Königlichen Gesellschaft der Wissenschaften zu Göttingen. In Commission bei B. G. Teubner in Leipzig. 1903, p. 299)

cylinder, the vectors $n(x)$ all lie on a circle on S^2, so that the image area, and thus the curvature, still equals zero. On a sphere of radius r, say centered at the origin the Gauss map is $G(x) = x/r$, so the curvature is constant and equal to $1/r^2$. On domains where the surface has a saddle, the Gauss map changes orientation, and thus the Gaussian curvature is negative.

Note that this definition of the Gaussian curvature depends on how S is embedded in $\mathbb{R}^3$. One of Gauss' key results is his "theorema egregium" (remarkable theorem) that states that this curvature depends only on the geometry of the surface S itself (all necessary measurements can be performed on S), but it does not depend on any embedding of S into an ambient space such as $\mathbb{R}^3$: *Gaussian curvature is an intrinsic quantity of any surface*. For instance, any smooth deformation of a sheet of paper (such as rolling up the paper) preserves distances measured within the paper, so the Gaussian curvature is zero as for a flat sheet of paper.

The famous *Gauss-Bonnet Theorem* says that for any closed oriented surface that is homeomorphic to a sphere, the integral of the Gaussian curvature over the surface equals 2π. More generally, suppose S is a closed oriented surface of Euler characteristic χ (with $\chi = 2$ for a sphere, $\chi = 0$ for a torus, $\chi = -2$ for a double torus, etc., each further "hole" adding -2 to the Euler characteristic; in general, if the surface S is triangulated by F triangles with E edges and V vertices, we have $\chi = V - E + F$). Then the integral of the Gaussian curvature over S is $\int_S \kappa(x)\,d^2x = 2\pi\chi(S)$. For a triangle Δ bounded by three geodesic segments, the sum of the three interior angles equals $\int_\Delta \kappa(x)\,d^2x + \pi$. In the special case of a sphere of radius r, we have $\kappa(x) \equiv \kappa = 1/r^2$; if δ denotes the sum of the interior angles, this implies $\delta - \pi = \kappa A = A/r^2$: this is the result mentioned above.

A combinatorial analog of the Gauss-Bonnet Theorem can be stated for polyhedra: for these, the Gauss map is constant on all faces, and all the curvature is concentrated in the vertices. The amount of curvature concentrated in a vertex v equals $2\pi - \delta(v)$, where $\delta(v)$ is the sum of all interior angles at v of all faces at v. This combinatorial version of the Gauss-Bonnet Theorem was the content of the mathematical lecture in the brief invitation movie to the IMO 2009 shown at the IMO 2008 in Madrid, and it is also what Gauss measured in his survey maps.

We would like to thank Malte Lackmann, Armin Leutbecher, and Peter Ullrich for useful suggestions for this text.

Chapter 4
Programme

4.1 IMO Programme Overview

Date	Leaders	Deputy Leaders	Participants
Friday **July 10**	Arrival		
Saturday **July 11**	Jury meeting		
Sunday **July 12**	Jury meeting		
Monday **July 13**	Jury meeting	Arrival	Arrival
Tuesday **July 14**	Opening ceremony		
Wednesday **July 15**	Questions & Answers	First day of contest	First day of contest
Thursday **July 16**	Questions & Answers Transfer to Bremen	Second day of contest	Second day of contest
Friday **July 17**	Coordination		Excursion
Saturday **July 18**	Coordination		Excursion
Sunday **July 19**	Last jury meeting		
	50th IMO anniversary celebration		
Monday **July 20**	Excursion		
Tuesday **July 21**	Closing ceremony Farewell dinner and party		
Wednesday **July 22**	Departure		

4.2 Programme Details

Friday, July 10

Leaders	
	Arrival
09:00	Short list handout
12:30	Lunch
	Arrival
19:00	Dinner

Saturday, July 11

Leaders	
07:00	Breakfast
09:00	Jury meeting
12:30	Lunch
16:00	Solutions handout
18:00	Jury meeting
19:00	Reception
20:30	Jury meeting

Sunday, July 12

Leaders	
07:00	Breakfast
09:00	Jury meeting
10:30	Coffee break
11:00	Jury meeting
12:30	Lunch
14:00	Jury meeting
15:30	Coffee break
16:00	Jury meeting
19:00	Dinner
20:30	Jury meeting

Monday, July 13

Leaders		Deputy Leaders and Contestants	
07:00	Breakfast		Arrival
09:00	Jury meeting		
10:30	Coffee break		
11:00	Translation of problems		
12:30	Lunch	13:00	Lunch
14:00	Jury meeting	14:30	Free time / Arrival
15:30	Coffee break		
16:00	Translation of problems		
19:00	Dinner	19:00	Dinner
20:30	Jury meeting	20:30	Free time

Tuesday, July 14

Leaders		Deputy Leaders and Contestants	
07:00	Breakfast	07:00	Breakfast
09:30	Bus departure	09:30	Bus departure
11:00	Opening ceremony		
14:00	Bus departure	15:00	Bus departure
14:30	Reception	15:30	Lunch
16:00	Bus departure	16:30	Free time
17:30	Jury meeting		
19:00	Dinner	19:00	Dinner
20:30	Jury meeting	20:30	Free time

Wednesday, July 15

Leaders		Deputy Leaders		Contestants	
07:00	Breakfast	07:00	Breakfast	06:30	Breakfast
				07:45	Bus departure
09:00	Q&A	09:00	Excursion	09:00	Exam 1
10:00	Coffee break			13:30	Bus departure
12:30	Lunch	14:00	Lunch	14:00	Lunch
		15:00	Free time	15:00	Free time
19:00	Dinner	19:00	Dinner	19:00	Dinner
		20:30	Free time	20:30	Free time

Thursday, July 16

Leaders		Deputy Leaders		Contestants	
07:00	Breakfast	07:00	Breakfast	06:30	Breakfast
		09:00	Free time	07:45	Bus departure
09:00	Q&A			09:00	Exam 2
11:00	Bus departure			13:30	Bus departure
14:00	Lunch	14:00	Lunch	14:00	Lunch
		15:00	Coordination	15:00	Free time
19:00	Dinner	19:00	Dinner	19:00	Dinner
		20:30	Coordination	20:30	Free time

Friday, July 17

Leaders and Deputy Leaders		Contestants	
07:00	Breakfast	07:00	Breakfast
09:00	Coordination	09:15	Bus departure
13:00	Lunch		Excursion with packed lunch
14:30	Coordination		
19:00	Dinner	19:00	Dinner
20:30	Coordination (optional)	20:30	Free time

Saturday, July 18

Leaders and Deputy Leaders		Contestants	
07:00	Breakfast	07:00	Breakfast
09:00	Coordination	09:15	Bus departure
13:00	Lunch		Excursion
14:30	Coordination	16:00	Barbeque
19:00	Dinner		
20:30	Coordination	20:30	Free time

Sunday, July 19

Leaders and Deputy Leaders		Contestants	
07:00	Breakfast	07:00	Breakfast
08:30	Jury meeting	09:00	Free time
11:30	Lunch		
13:00	Bus departure		
14:30	50th IMO anniversary celebration		
19:00	Bus departure		
20:30	Dinner		
21:30	Free time		

Monday, July 20

Leaders, Deputy Leaders and Contestants	
06:30	Breakfast
08:00	Bus departure
	Excursion to Wangerooge island full day of activities including lunch and supper
22:00	Bus arrival at Jacobs University

Tuesday, July 21

Leaders, Deputy Leaders and Contestants	
07:00	Breakfast
09:00	Bus departure
10:30	Awards and Closing Ceremony
13:30	Bus departure
14:30	Lunch
15:30	Free time
20:00	Farewell party

Wednesday, July 22

Leaders, Deputy Leaders and Contestants	
07:00	Breakfast
	Departure

4.3 Daily News

Each day we issued the "IMO Daily News" to facilitate communication between the 565 contestants and their team guides and deputy leaders, and also with the jury and the coordinators, once they moved to the Jacobs campus. In all, we had to keep more than 1000 people updated on what was going on. The Daily News was produced in three stages: first the editorial content was assembled by Wilfried Kurth and Stephanie Schiemann; then in the middle of the night the information was put together in an attractive format by Marcel Oliver; and finally it was printed and distributed early in the morning by Anke Allner. Reading the Daily News again after the IMO brings back lively memories of our exciting time in Bremen.

𝔇𝔞𝔦𝔩𝔶 𝔑𝔢𝔴𝔰

DAY 1   *JULY 13, 2009*   BREMEN, GERMANY

Arrival Day

Long before breakfast, the first teams for the 50th International Mathematics Olympiad were welcomed by their team guides at Bremen Airport or at the train station and shuttled to the Jacobs University campus for check-in.

By late afternoon, over 40 teams had arrived. So far, despite the unfortunate cancellation of the team of Brunei, as well as continuing uncertainty about the participation of Bangladesh, the 50th Olympiad is its way to be the biggest ever. We are particulary happy to report the arrival of the teams of Honduras and Mauritania, who were initially thought missing.

This year's participants come from more than 100 countries, the youngest contestant at the age of 11. In addition, there are 150 guides and assistants, more than 60 coordinators, 40 invigilators, an office staff of 15, and an organizing committee of 30 on the ground. It is now the second time the IMO is held in Germany; the first German IMO took place 1989 in Braunschweig.

After a wet Sunday, many participants have been enjoying the sunny skies playing beach volleyball, tennis, basketball, soccer, and other outdoor sports. For rainier days, the colleges also feature indoor table tennis, pool, and table soccer.

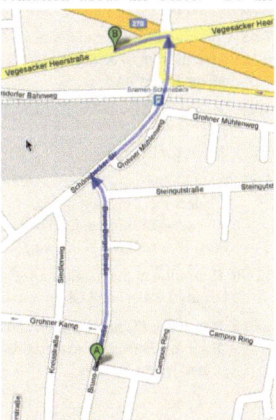

So far, all teams arrived safe and sound, except for minor delays and confusion about the buses. So the important questions of the day were "Where do I find a bank?" (there is an ATM in the post office on Vegesacker Heerstraße, see map), "Where can I find a supermarket?" (there is a supermarket in the *Haven Höövt* shopping mall near Vegesack harbor and Vegesack train station, see bottom map), "Where is the internet café?" (in the ground floor of the building *Research I*, follow the posted signs), "Where can I buy a phone card?" (at the bistro "Friseur" next to the porter's lodge at the Jacobs main entrance), "Where can I buy a mobile phone?" (mobile phones are not allowed inside of the contest hall...), and "Can I get a second key to my room?" (huh???).

From the main gate to the post office

On Tuesday

7:00 Breakfast

7:45 Team Guides depart for Pier 2, meet in front of Research I

9:15 Contestants depart for Pier 2, meet in front of Research I

11:00 Opening ceremony at Pier 2

After ceremony Group photograph at Pier 2

Back on campus Team photographs in the Campus Center

WEATHER REPORT

Tuesday will be cloudy with some showers, clear in the evening. Wednesday and Thursday mostly sunny with a small chance of rain. Thunderstorms on Friday. High temperatures 25 °C.

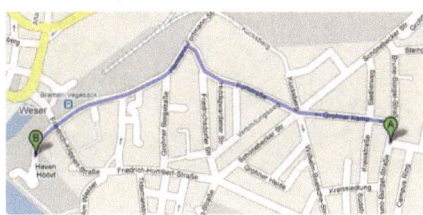

From the main gate to the Haven Höövt shopping mall

Daily News

DAY 2   *JULY 14, 2009*   BREMEN, GERMANY

Opening Ceremony

Today we had a wonderful opening ceremony at Pier 2 on the river Weser. The team guides had already gotten up very early in the morning to practice their country's parade.

The hall at Pier 2 has two-tiered seating: jury and observers took seat on the upper balcony, while the teams themselves were seated in the main hall. This way, everybody could watch and enjoy, but no direct contact between jury and teams was possible.

The opening event was chaired by Prof. Dr. Beutelspacher, Di-
rector of the
Mathematicum
in Gießen, a
museum of
Mathematics.
After an expres-
sive breakdance
performance
by *Breakma-
trix*, we heard

a number of greetings, including one from German chancellor Dr. Angela Merkel, physician and former contestant of the national Mathematics Olympiad in Eastern Germany. Prof.

Beutelspacher gave a short and short-whiled mathematical interlude "Calculating without Calculator". Then Andreas Storm, Parliamentary State Secretary to the Federal Minister of Education and Research, opened the 50th International Mathematics Olympiad 2009 in Bremen, the first ever with more than 100 teams.

The highlight, of course, was the parade of the IMO 2009 teams.

On Wednesday

6:30 Breakfast

7:45 Contestants and Team Guides meet in front of Research I, depart for Exam 1

8:45 Deputy Leaders and Observers B meet in front of Research I, depart for excursion

9:45 Team Guides and Deputy Leaders meet in front of the Übersee-Museum, guided by Prof. Ronny Wells

12:00 Reception at the Schütting for Deputy Leaders and Observers B

13:15 Return for Deputy Leaders and Observers B from Wachtstraße, guided by Prof. Wells

from 13:00 Team Guides and Contestants return to their buses

Back on campus Lunch

Important information for contestants: do not forget to check out at the welcome desk whenever you leave the campus. You may venture out together with contestants of other teams, but you must be accompanied by one team guide.

We wish all contestants a successful competition!

WEATHER REPORT

Now that today's weather already beat the weather forecast, tomorrow's going to be mostly sunny with a high of 26 °C. Thursday clear skies; rain or thunderstorms on Friday.

CORRECTION

We are embarrassed to admit, but yesterday's claim that this is Germany's second IMO was badly mistaken. It is in fact the fourth, after Berlin 1965 and Erfurt 1974 in the former German Democratic Republic. Apologies!

Opening Ceremony at Pier 2

The Parade

Daily News

DAY 3   *JULY 15, 2009*   BREMEN, GERMANY

First Exam

The third day – the first exam. While deputy leaders and observers B and C enjoyed a guided tour through the city of Bremen on a perfect summer day, contestants were sweating over math and invigilators walking their walk back and forth the examination hall. The estimation of the success in solving the problems is an individual thing. At least the organizers, who were almost as anxious about the exam, are greatly relieved that, despite the scale of the event, the logistics worked out very well – and this is important

And after the contest, meeting people from many other parts of the world, everyone got some relief on the campus lawn or in the common room – and this is equally important, of course.

On Thursday

6:30 Breakfast

7:45 Contestants and Team Guides meet in front of Research I, depart for Exam 2

from 15:00 Soccer competition

20:00 Informal lecuture "Fluids, Math, and a Millenium Challenge" by Marcel Oliver, Jacobs University, in the Research II Lecture Hall

There is no special program for Deputies and Observers.

We wish all contestants a successful competition, especially to those who do not feel so happy about today!

WEATHER REPORT

Tomorrow will be clear, high temperature 26 °C. Rain or thunderstorms on Friday with temperatures up to 29 °C; the forecast for the weekend is mixed.

IN THE PRESS

Coverage of the opening ceremony by local TV station Radio Bremen, including a short video clip in German:

`http://tinyurl.com/lr2yuo`

Inside the hall

Outside the hall after the exam

Message from the Chief Invigilator

By DIERK SCHLEICHER

I hope you share my view that the first contest day went quite smoothly. I greatly appreciated your cooperation. Nonetheless, there are a few issues how you could help yourself and us. Please read carefully.

- Please reuse the transparent folders for bringing your pens, rulers, and other solution items into the contest rooms. There is no separate set for the second contest day.

- On question/answer forms, please fill in your contestant code and seat number: otherwise, we will not be able to deliver the answer of the jury back to you. (And write the question on the sheet we provided, not on the yellow question card!)

- Also remember to write your contestant code and problem number onto all the solution sheets that you put into the colored cardboard folders. (And don't write your solutions on the colored cardboard folders!)

- The official IMO paper may also be used as scrap paper – the jury evaluates only those sheets that are returned within the colored cardboard folders!

- Instructions on toilet visits: if you need to go to the toilet before the contest, do that before departing from the Jacobs campus or at before entering the contest hall. Once you pass through the entrance doors to the contest room, we ask you not to leave the room again before the contest starts. During the contest, once you raised the green "toilet card", wait at your place until an invigilator comes to you.

- On the way to and from the toilet, use the wide alleys; do not pass between tables.

- Finally, we pay a deposit on the water bottles. Return them to us; don't discard them!

I would like to express my sincere gratitude to the wonderful team of invigilators! Let us make the second contest day equally smooth and efficient as today!

Also in the name of my colleagues, I wish you a successful second contest day!

Daily News

Second Exam

Exams are over! Everybody has done their best, and the contestants are now busy enjoying the social program.

As the organizing team, we would like to say *thank you* to all hardworking people who helped during the contest days.

We welcome the jury, the committee, and all coordinators who arrived today on the Jacobs campus from their secret hideout in Bremerhaven. Contestants, please

Exam scanning   respect that the jury-zone, including College Nordmetall, is off-limits for you.

In the afternoon, we saw the kick-off for the IMO soccer tournament, with multinational teams crossing all sorts of political and geographical boundaries. There is, for instance, a joint team of North and South Korea together with Vietnam and Pakistan, another between Serbia, Bosnia and Herzegovina, and Montenegro, or Cyprus, Greece, Ecuador and Venezuela. Or Poland, Norway, Azerbaijan and Mongolia. Watch this space for detailed coverage of the quarter- and semi-finals which take place on Saturday.

On Friday

Tomorrow will be the first excursion day for the contestants and team guides. They will be accompanied by Observers C, the senior and chief guides.

7:00 Breakfast

from 7:00 Team Guides pick up lunch packages for their teams

8:00 Departure Bus 11

9:00 Departure Buses 5, 6, 13, 14

9:30 Departure Buses 1, 2, 3, 4, 12

10:00 Departure Buses 7, 8, 9, 10

19:00–22:00 Gym open

20:00–22:00 Board Game Lounge / The Other Side

20:30–23:00 IMO-Cinema: *Good will hunting* by Gus van Sant

21:00 Informal Math Lecture "Newton's Root-Finding Method, Chaos Theory, and Unsolved Questions" by Dierk Schleicher, Jacobs University, in the Research II Lecture Hall

We wish you all a wonderful excursion day and successful work for the jury.

WEATHER REPORT

Rain or thunderstorms on Friday with temperatures up to 25 °C; mixed weather with temperatures around 20 °C from Saturday.

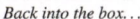

Back into the box...

...and onto the lawn.

Daily News

DAY 5   *JULY 17, 2009*   BREMEN, GERMANY

Excursions 1

While the jury is working, the contestants had their first excursion day, with groups going to

Meyer-Shipyard Papenburg,

Transrapid maglev track Lathen,

Miniatur Wunderland Hamburg,

and Klimahaus Bremerhaven.

On Saturday

from 7:00 Breakfast

from 7:00 Team Guides pick up lunch packages at the IRC

9:30 Departure Buses 4, 9, 5, 10, 11 to Sportsgarden

9:30 Departure Buses 3, 6, 7, 8, 12, 13, 14 to Schlachte

9:45 Departure Buses 1, 2 to Eurogate Bremerhaven (please take your ID card or passport along)

15:00 Arrival back on campus

from 15:30 Barbecue

16:00 Team and senior guide meeting in the CNLH

19:30–21:00 Soccer quarter- and semifinals

20:00 Informal Math Lecture "How to solve diophantine equations" by Michael Stoll, Universität Bayreuth, in the Research II Lecture Hall

20:00–22:00 Board Game Lounge / The Other Side

22:30–01:30 IMO-Disco with Karaoke Competition / The Other Side

Enjoy the day!

SOCCER QUARTERFINALS

Finland/India/Romania/Turkmenistan (1A) – Iran/Netherlands/Denmark (2B)

Slovenia/Croatia/Argentina/Albania (2A) – Brazil/Mexico/Honduras/Chile (1B)

Ireland/Georgia/Bangladesh/Zimbabwe (1C) – Poland/Norway/Azerbaijan/Mongolia (2D)

Ecuador/Greece/Cyprus/Venezuela (2C) – Serbia/Bosnia&Herzegovina/Montenegro (1D)

SHORT ANNOUNCEMENTS

The common-rooms in the Colleges are still closed.

Thai Le Tran, team guide of Vietnam, will organize an outdoor game – a mix of dodgeball and relay race – at 9:30 on Sunday on the Campus green. Rules will be explained by him just then. Participants should wear colored T-shirts. Some guides will be needed as judges, please contact him at 0176-49415196.

We have received this greeting, which we will relay here: *Soy el Dr. Marcelo Abad, de Ecuador. Les deseo exitos a los organizadores y participantes de la IMO 2009, y en especial a mi hijo Christian Abad Coronel representante de Ecuador. Saludos, Dr. Abad*

WEATHER REPORT

Partially cloudy in the morning with possibly a few rain showers in the afternoon. Sunday and Monday some rain, chance of thunderstorms. Temperatures around 20 °C. Tuesday sunny and up to 25 °C.

𝕯𝖆𝖎𝖑𝖞 𝕹𝖊𝖜𝖘

DAY 6   *JULY 18, 2009*   BREMEN, GERMANY

Excursions 2 and Barbecue

Fortunately the weather on Saturday was better than forecast. The contestants enjoyed excursions to Bremen city and the *Schlachte* with its flea market. Others went for a challenge in the sports garden or walked around the soccer stadium. Yet another group went to visit the Eurogate container port in Bremerhaven.

chance to improve is on Sunday when buses for all participants must depart at 13:00 sharp for the 50th anniversary celebration at the Musical Theater Bremen.

In the afternoon, despite the windy weather, everybody enjoyed the barbecue in front of the Campus Center.

Unfortunately, some buses couldn't start on time in the morning because many contestants were not at the meeting point on schedule. This makes the work of the guides needlessly harder and should be changed. The next

On Monday

from 6:00 Breakfast

6:45 Contestants, team guides and senior guides of buses 1, 2, 3, 4, 7, 8, 12 and travelers of buses A, B, C, D, E, F (Leaders, Deputies, Observers A and B) and buses G and H (transporting honorary guests and coordinators) meet for departure to Wangerooge

7:00 contestants, team guides and senior guides of buses 5, 6, 9, 10, 11, 13, 14 meet for departure to Wangerooge

WEATHER REPORT

Sunday and Monday some rain, chance of thunderstorms. Temperatures around 20 °C. Tuesday sunny and up to 25 °C. Unstable weather conditions for the rest of the week.

SHORT ANNOUNCEMENTS

Especially on Monday, everybody must be at the buses on time. We will otherwise miss the ferry which can't wait due to the falling tide. It won't be fun to just glimpse at the island across the mudflat. And our lunch is there! Since we cannot extend the opening hours for breakfast, Aramark will provide bags (without content) in the evening or and in the morning so that people can carry some food along. The team guides should have breakfast together with their teams.

*

On Sunday

from 7:00 Breakfast

9:00–12:00 and 20:00–22:00 Gym open

9:30 Field game on the Campus Green

from 10:00 Photo session, please check the schedule posted in the IRC entrance hall

13:00 Departure for the 50th IMO anniversary celebration

20:00–22:00 Soccer tournament final

20:00–22:00 Board Game Lounge / The Other Side

DAY 6   **Daily News** *JULY 18, 2009*   2

To avoid misunderstandings, please note that the senior guides are responsible for the buses and their decisions have to be followed. Since the question came up, we repeat here again: outside of the Jacobs campus the team members have to stay with their team guides. Observers C may venture out on their own.

*

Remember that on Sunday at 9:30 the field game takes place on the Campus Green (the lawn amidst the research buildings). The more participants come, the better! The rules in short:

1. There will be four teams, based on their shirt color.
2. Carry the Flag of Bremen to the corners of the Green to score.
3. Shoot other players with tennis balls to claim the flag for your team, and avoid to get shot.

Everything else will be explained at the site. Bring drinking water. The game will last for about one hour.

*

A wonderful IMO-story: Eduardo Pérez Almarales, the only contestant of Cuba meets his grandmother, who came traveling from Madrid to Bremen to visit him, on campus.

SOCCER QUARTERFINALS

Finland/India/Romania/Turkmenistan – Brazil/Mexico/Honduras/Chile	0:4
Slovenia/Croatia/Argentina/Albania – Iran/Netherlands/Denmark	3:2
Ireland/Georgia/Bangladesh/Zimbabwe – Poland/Norway/Azerbaijan/Mongolia	3:1
Ecuador/Greece/Cyprus/Venezuela – Serbia/Bosnia and Herzegovina/Montenegro	0:1

SOCCER SEMIFINALS

Semifinals: Brazil/Mexico/Honduras/Chile – Slovenia/Croatia/Argentina/Albania	1:0
Serbia/Bosnia and Herzegovina/Montenegro – Ireland/Georgia/Bangladesh/Zimbabwe	0:2

So in the final, Ireland/Georgia/Bangladesh/Zimbabwe will play Brazil/Mexico/Honduras/Chile. Kick-off Sunday, 20:00.

Flea market at the Schlachte   *At the barbecue*

𝕯𝖆𝖎𝖑𝖞 𝕹𝖊𝖜𝖘

DAY 7   *JULY 19, 2009*   BREMEN, GERMANY

50th Anniversary

The celebration of the 50th Anniversary of the International Mathematical Olympiad was a wonderful event. The young participants of the IMO had the chance to hear about the research and the IMO experiences of 6 of the most famous mathematicians of the world. They also had the possibility to talk to them in the intermissions, where some guests of honor were nearly crushed by the crowds...

We hope that this event will cheer many of the contestants to a successful mathematical career so that, possibly, one will be a guest of honor at the 75th or 100th IMO anniversary.

On Monday

from 6:00 Breakfast

6:45 Contestants, team guides and senior guides of buses 1, 2, 3, 4, 7, 8, 12 and travelers of buses A, B, C, D, E, F (Leaders, Deputies, Observers A and B) and buses G and H (transporting honorary guests and coordinators) meet for departure to Wangerooge

7:00 contestants, team guides and senior guides of buses 5, 6, 9, 10, 11, 13, 14 meet for departure to Wangerooge

Have a wonderful and sunny day on Wangerooge!

WEATHER REPORT

Monday with some rain, chance of thunderstorms. Temperatures around 20 °C. Tuesday sunny and up to 25 °C. Unstable weather conditions for the rest of the week.

SHORT ANNOUNCEMENTS

Some last information for our trip to the island Wangerooge: Team guides must make sure that their team is back at the ferry in time for return departure. Please meet 15 minutes before the announced times. Be careful not to walk too far into the the mudflats because the tide can rise very quickly. Respect that the island is also a nature reserve; use the garbage bins!

*

Team guides and senior guides meet directly after the trip in CNLH.

*

Check out procedure for people leaving before July 22:
• Up to 19:00: Please hand in keys in the IMO office in the Campus Center,
• After 19:00: Please hand in keys at the porter's lodge.
People leaving on 22 July also check out at the porter's lodge.

*

All meals are available on Wednesday.

*

Team guides should check that all their team members have their keys before leaving their rooms on 22 July. Anyone loosing their key will be charged 50 Euros.

*

Remember to vote for Miss and Mr. IMO until 23:59 on Monday.

*

And the winners of the soccer tournament are Brazil/Mexico/Honduras/Chile, who beat Ireland/Georgia/Bangladesh/Zimbabwe 1:0.

Intermission at the Bremen Musical Hall

The winners of the soccer tournament

Daily News

DAY 8   *JULY 20, 2009*   BREMEN, GERMANY

Wangerooge

When we got onto the ferry boats at Harlesiel in the morning, we were greeted by strong winds, rain, and it was cold—about 12 °C. On the island train it was pouring. So who expected that we would have a nice, sunny afternoon on the beach? But we did, with lots of fun besides.

Chief Invigilator

Keeping the balance

On Tuesday

7:00 Breakfast

8:15 Team guides meet in front of Research I and depart for the Glocke at 8:30

9:15 Contestants, deputies and senior guides meet in front of Research

I and depart to the Glocke at 9:30 (buses 1–14)

10:00 Departure for all others (buses A–H)

11:00 Awards and Closing Ceremony in the Glocke; after the ceremony guided free time in the city of Bremen

16:00–17:00 Interested in studying at Jacobs? – The admissions team provides information in the IRC (Campus Center)

19:00 Team and senior guide meeting

20:30 Farewell Party

Gold medal winner

WEATHER REPORT

Sunny and up to 25 °C. Unstable weather conditions for the rest of the week.

SHORT ANNOUNCEMENTS

Check out procedure for people leaving before July 22:
• Up to 19:00: Please hand in keys in

the IMO office in the Campus Center,
• After 19:00: Please hand in keys at the porter's lodge.
People leaving on 22 July also check out at the porter's lodge.

∗

All meals are available on Wednesday.

∗

Team guides should check that all their team members have their keys before leaving their rooms on 22 July. Anyone loosing their key will be charged 50 Euros.

∗

A mobile telephone and a jacket was found in the Musical Theater. Please contact the office.

∗

We keep a photo collection of the 2009 IMO on a German web site for young math talents:

`http://tinyurl.com/ne3ypn`

If you register, you may upload your own pictures or exchange messages with other participants. The user interface is currently German only, but we will try to set up an English version if technically possible.

Two copies of $\mathbb{S}^1$

Down onto the beach...

...beach closed;

high above the water...

...high in the air.

Proving theorems...

...all the way to victory.

𝔇𝔞𝔦𝔩𝔶 𝔑𝔢𝔴𝔰

DAY 9 | *JULY 21, 2009* | BREMEN, GERMANY

Awards and Closing Ceremony

From the Award and Closing Ceremony we can tell you that is was an impressive celebration. But we do not need to tell you, because you were there. We only can congratulate all of you to your achievement, equal if you won a medal or not. You can be proud!

This is our last Daily News issue for the 50th IMO 2009. Normally they were written late at night, somewhere 11 pm and 4 am. This time we are trying to get some rest, so we cannot yet tell you whether the farewell party was wonderful, but we are sure it will be.

TOP PARTICIPANTS

We congratulate all participants to their achievements. Very impressive scores were achieved by the three top participants.

Dongyi Wei, People's Republic of China, first participated at the IMO in 2008. Both times he achieved gold with a perfect score.

Makoto Soejima, Japan, had two prior IMO participations, where he won bronze 2007 and gold 2008. This year he achieved gold with a perfect score.

Lisa Sauermann, Germany, first won a silver medal at the IMO 2007 and gold medals 2008 and 2009, this year only one point away from a perfect score.

We also note that the three best buses were Bus 8 (CHN, HKG, JPN, KOR, MAC, PRK, TWN) with an average score of 27.14 coming in first, Bus 2 (BLR, GEO, KAZ, MDA, MNG, RUS, UKR) with 22.71 in second place and Bus 7 (IDN, IRN, KHM, MAS, THA, VNM) with 19.75 third .

GUESTS OF HONOR

We say *thank you* to the six main speakers of Sunday's anniversary celebration who took the time to be present at the award ceremony to present the medals to the successful participants.

Béla Bollobás, University of Cambridge and University of Memphis, works in functional analysis, combinatorics, graph theory and percolation. Born

1943, he won a bronze and two gold medals at the first ever IMOs 1959–1961.

Timothy Gowers, Cambridge University, works in combinatorics and functional analysis. Born 1963, he won an IMO gold medal in 1981 and the

Fields medal in 1998.

László Lovász, Eötvös Loránd University Budapest, works in discrete mathematics and theoretical computer science. Born 1948, he won three IMO

gold and one silver medal 1963–1966. He is currently the president of the International Mathematical Union.

Stanislav Smirnov, University of Geneva, works in complex and geometric analysis, dynamical systems, probability theory and percolation. Born 1970, he

won two IMO gold medals in 1986 and 1987.

Terence Tao, University of California at Los Angeles, works in harmonic analysis, partial differential equations, probability, ergodic theory, combi-

natorics, compressed sensing, representation theory, and number theory. Born 1975, he won bronze, silver, and gold at the IMO 1986–1988 and the Fields medal in 2006.

Jean-Christophe Yoccoz, Université Paris XI, Orsay, works in dynamical systems, complex analysis, and number theory. Born in 1957, he won IMO

silver and gold medals in 1973 and 1974, and the Fields medal in 1994.

SHORT ANNOUNCEMENTS

Changed check-out procedure Please check out at the IMO office in the Campus Center. If the office is closed, you may check out at the porter's lodge.

*

All meals are available on Wednesday.

*

Team guides should check that all their team members have their keys before leaving their rooms on 22 July. Anyone loosing their key will be charged 50 Euros.

*

We keep a photo collection of the 2009 IMO on a German web site for young math talents:

`http://tinyurl.com/ne3ypn`

If you register, you may upload your own pictures or exchange messages with other participants. The user interface is currently German only, but we will try to set up an English version if technically possible.

*

We thank all of you for coming to Bremen and wish you a good return to your home. We look forward to seeing you again at future IMOs, at Jacobs University, and as active members of the international mathematics community.

*

You can find an archive of the daily news letters (with color pictures) on

`http://tinyurl.com/ls826o`

*

Daily News was brought to you by Stephanie Schiemann, Marcel Oliver and Wilfried Kurth.

At the closing ceremony

The Korean teams

Chapter 5
Participants

5.1 National Teams

ALB   Albania

Leader	Fatmir Hoxha
Deputy	Artur Baxhaku
Contestants	Andi Reçi, Tedi Aliaj, Ornela Xhelili, Arlind Gjoka, Niko Kaso, Ridgers Mema

ALG   Algeria

Leader	Abed-Seddik Bouchoucha
Deputy	Ali Atia
Contestants	Lamia Attouche, Hacen Zelaci, Oussama Guessab, Kacem Hariz,

ARG   Argentina

Leader	Patricia Fauring
Deputy	Flora Gutierrez
Contestants	Germán Stefanich, Alfredo Umfurer, Iván Sadofschi, Federico Cogorno, Ariel Zylber, Miguel Maurizio

ARM   Armenia

Leader	Nairi Sedrakyan
Deputy	Koryun Arakelyan
Contestants	Anna Srapionyan, Nerses Srapionyan, Vahagn Kirakosyan, Vahagn Aslanyan, Hayk Saribekyan, Vanik Tadevosyan

 **ALB**

  **ALG**

  **ARG**

  **ARM**

AUS   Australia

Leader   Angelo Di Pasquale
Deputy   Ivan Guo
Contestants   Aaron Wan Yau Chong, Andrew Elvey Price, Stacey Wing Chee Law, Alfred
                Liang, Dana Ma, Sampson Wong
Observer A   Hans Lausch, Ian Roberts

AUT   Austria

Leader   Robert Geretschläger
Deputy   Walther Janous
Contestants   Felix Dräxler, Adrian Fuchs, Johannes Hafner, Clemens Müllner, Stephan
                Pfannerer, Valerie Roitner

AZE   Azerbaijan

Leader   Fuad Garayev
Deputy   Jafar Jafarov
Contestants   Ilgar Ramazanli, Elvin Aliyev, Zulfu Aslanli, Altun Shukurlu, Subhan Rus-
                tamli, Emil Jafarli

BGD   Bangladesh

Leader   Mahbub Alam Majumdar
Deputy   A A Munir Hasan
Contestants   Samin Riasat, Nazia Naser Chowdhury, Tarik Adnan Moon, Haque Muhammad
                Ishfaq, Kazi Hasan Zubaer, Pranon Rahman Khan
Observer C   Jesmin Akter, Md. Ibrahim Khalilullah

AUS

AUT

AZE

BGD

BLR   Belarus

Leader   Igor Voronovich
Deputy   Sergei Mazanik
Contestants   Sviatlana Auchynnikava,   Sergey Dovgal,   Artsiom Hovarau,   Palina Khudzi-
                akova, Artsiom Toyestseu, Aliaksei Vaidzelevich

BEL   Belgium

Leader   Bart Windels
Deputy   Philippe Niederkorn
Contestants   Loïc Burger, Cédric De Groote, Nicolas Radu, Gregory Debruyne, Alexander
                Lemmens, Mats Vermeeren
Observer A   Ria Van Huffel

BEN   Benin

Leader   Assogba Bernardin Kpamegan
Deputy   Emile Sacla
Contestants   Arélyss Eblohoué, Comlan Edmond Koudjinan

BOL   Bolivia

Leader   René Arturo Aguilar Vera
Contestants   Erick Daniel Vicente Minaya,   Diego Salazar Gutiérrez,   Álvaro Rubén Hurtado
                Maldonado

 **BLR**

 **BEL**

 **BEN**

 **BOL**

BIH   Bosnia and Herzegovina

Leader   Damir Hasić
Deputy   Boris Marjanović
Contestants   Adnan Ademović, Admir Beširević, Jelena Radović, Vlado Uljarević, Ratko
              Darda, Mina Ferizbegović

BRA   Brazil

Leader   Carlos Yuzo Shine
Deputy   Ralph Costa Teixeira
Contestants   Henrique Pondé de Oliveira Pinto, Renan Henrique Finder, Marcelo Tadeu de Sá
              Oliveira Sales, Davi Lopes Alves De Medeiros, Marco Antonio Lopes Pedroso,
              Matheus Secco Torres da Silva

BGR   Bulgaria

Leader   Nikolai Nikolov
Deputy   Peter Boyvalenkov
Contestants   Lyuboslav Panchev, Svetozar Stankov, Zhivko Zhechev, Galin Statev, Sve-
              toslav Karaivanov, Victor Valov
Observer A   Emil Kolev

KHM   Cambodia

Leader   Lin Sok
Deputy   Sovann Suon
Contestants   Sopheak Touch, Kunthea Din, Leanghour Lim, Guechlaing Chea, Panha Ouk,
              Nam Seang

 **BIH**

 **BRA**

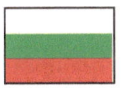 **BGR**

 **KHM**

CAN   Canada

Leader	Dorette Pronk
Deputy	David Arthur
Contestants	Robin Cheng, Jonathan Schneider, Xiaolin (Danny) Shi, Hunter Spink, Chen Sun, Chengyue (Jarno) Sun
Observer B	Jacob Tsimerman

CHI   Chile

Leader	Hernan Burgos
Deputy	Victor Cortes
Contestants	Anibal Velozo, Benjamin Baeza, Sebastian Zuñiga, Mauricio Garcia

CHN   People's Republic of China

Leader	Hua-Wei Zhu
Deputy	Gang-Song Leng
Contestants	Dongyi Wei, Bo Lin, Fan Zheng, Zhiwei Zheng, Yanlin Zhao, Jiaoyang Huang
Observer A	Bin Xiong
Observer B	Yunhao Fu

COL   Colombia

Leader	Maria Elizabeth Losada
Deputy	Jonathan Montaño
Contestants	Jorge Alberto Olarte, Angela Castañeda, Hayden Liu Weng, Jorge Francisco Barreras Cortes, Nicolás Del Castillo, Carlos Carvajal

CAN

CHI

CHN

COL

CRI Costa Rica

Leader   Mario Alberto Marín Sánchez
Deputy   Daniel Campos Salas
Contestants   Christopher Antonio Trejos Castillo, Rafael Ángel Rodríguez Arguedas, Anthony
                Santiago Chaves Aguilar, Ezequiel Heredia Fernández

HRV Croatia

Leader   Mea Bombardelli
Deputy   Matija Bašić
Contestants   Ivo Božić, Borna Cicvarić, Nina Kamčev, Adrian Satja Kurdija, Matko Ljulj,
                Goran Žužić

CUB Cuba

Leader   Eduardo Pérez Almarales
Contestants   Reynaldo Gil Pons

CYP Cyprus

Leader   Andreas Skotinos
Deputy   Savvas Timotheou
Contestants   Michael Anastos, Christos Anastassiades, Georgios Panagopoulos, Talia Tseri-
                otou, Maria Michaelidou, Neofytos Apostolou

CRI

HRV

CUB

CYP

CZE   Czech Republic

Leader   Jaroslav Švrček
Deputy   Martin Panák
Contestants   David Klaška, Jan Matějka, Josef Ondřej, Samuel Říha, Josef Tkadlec, Jan
              Vaňhara

DEN   Denmark

Leader   Jens-Søren Kjær Andersen
Deputy   Sune Precht Reeh
Contestants   Mathias Tejs Knudsen, Nikolaj Kammersgaard, Ben Braithwaite, Niels Olsen,
              Rasmus Nørtoft Johansen, Pernille Hanehøj
Observer B   Sven Toft Jensen

ECU   Ecuador

Leader   Jorge Chamaidan
Contestants   Xavier Soriano, Gabriel Bravo, Christian Abad, Miguel Ordoñez, Christian
              Pihuave, Caril Martinez

EST   Estonia

Leader   Härmel Nestra
Deputy   Oleg Košik
Contestants   Kairi Kangro, Heino Soo, Paavo Parmas, Rauno Siinmaa, Aleksandr Šved,
              Andre Tamm

CZE

DEN

ECU

EST

FIN   Finland

Leader   Matti Lehtinen
Deputy   Niko Vuokko
Contestants   Aleksis Koski, Konsta Lensu, Heikki Pulkkinen, Alexey Sofiev, Topi Talvitie,
              Lasse Vekama
Observer A   Maisa Spangar

FRA   France

Leader   Claude Deschamps
Deputy   Pierre Bornsztein
Contestants   Thomas Budzinski, Jean Garcin, Ambroise Marigot, Jean-François Martin,
              Sébastien Miquel, Sergio Véga
Observer A   Johan Yebbou

GEO   Georgia

Leader   Larry Gogoladze
Deputy   George Bareladze
Contestants   Beka Ergemlidze, Lasha Lakirbaia, Nikoloz Machavariani, Lasha Peradze,
              Tsotne Tabidze, Levan Varamashvili
Observer A   Amiran Ambroladze

GER   Germany

Leader   Eric Müller
Deputy   Thomas Kalinowski
Contestants   Bertram Arnold, Christoph Kröner, Malte Lackmann, Martin Merker, Jens
              Reinhold, Lisa Sauermann

FIN

FRA

GEO

GER

HEL   Greece

Leader	Anargyros Fellouris
Deputy	Evangelos Zotos
Contestants	Ilias Giechaskiel, Fotis Logothetis, Dimitrios Papadimitriou, Pappelis Konstantinos, Evangelos Taratoris, Ilias Zadik
Observer B	Georgios Apostolopoulos

GTM   Guatemala

Leader	Javier Ronquillo
Deputy	José García
Contestants	Francisco José Martínez Figueroa, Marcos Fernando Galindo Gomez, Alejandro José Vargas De León, José Carlos Arandi Ayala

HND   Honduras

Leader	Mario Roberto Canales Villanueva
Deputy	Juan Carlos Iglesias
Contestants	Nestor Alejandro Bermudez, Sergio David Manzanarez, José Ramón Madrid Padilla

HKG   Hong Kong

Leader	Tat Wing Leung
Deputy	Yuk Kam Lau
Contestants	Tak Wing Ching, Ping Ngai Chung, Cho Ho Lam, Ka Kin Kenneth Hung, Ngai Fung Ng, Tak Hei Yu
Observer B	Kin Sum Lee

HEL

GTM

HND

HKG

HUN   Hungary

Leader   József Pelikán
Deputy   Sándor Dobos
Contestants   András Éles, Kristóf Kornis, Dániel Nagy, János Nagy, Gergely Szűcs, István
               Tomon

ISL   Iceland

Leader   Auðun Sæmundsson
Deputy   Ragna Briem
Contestants   Arna Pálsdóttir, Helga Kristín Ólafsdóttir, Ögmundur Eiríksson, Helgi Kristjáns-
               son, Ingólfur Eðvarðsson, Paul Frigge

IND   India

Leader   Chudamani R Pranesachar
Deputy   Zafar Ahmed
Contestants   Akashnil Dutta, Akshay Mittal, Gaurav Digambar Patil, Ananth Shankar,
               Sameer Wagh, Subhadip Chowdhury
Observer B   Shashikant Anant Katre

IDN   Indonesia

Leader   Hery Susanto
Deputy   Yudi Satria
Contestants   Joseph Andreas, Aldrian Obaja Muis, Andreas Dwi Maryanto Gunawan, Raja
               Oktovin Parhasian Damanik, Ivan Wangsa Cipta Lingga, Ronald Widjojo
Observer A   Budi Surodjo
Observer B   Purwadi Sutanto, Surya Wijaya

HUN

ISL

IND

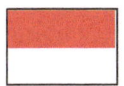

IDN

IRN   Islamic Republic of Iran

Leader	Arash Rastegar
Deputy	Omid Naghshineh
Contestants	Seyed Ehsan Azarmsa, Hossein Dabirian, Amirmasoud Geevechi, Sepehr Ghazi Nezami, Khashayar Khosravi, Pooya Vahidi Ferdowsi
Observer A	Omid Hatami, Mohsen Jamaali
Observer B	Aliakbar Daemi

IRL   Ireland

Leader	Bernd Kreussler
Deputy	Gordon Lessells
Contestants	Jack McKenna, David McCarthy, Colman Humphrey, Colin Egan, Vicki McAvinue, Cillian Power
Observer A	Donal Hurley

ISR   Israel

Leader	Shay Gueron
Deputy	Eran Assaf
Contestants	Omri Ben Eliezer, Ofir Gorodetsky, Ohad Nir, Inbar Klang, Rom Dudkiewicz, Yuval Goldberg

ITA   Italy

Leader	Roberto Dvornicich
Deputy	Francesco Morandin
Contestants	Fabio Bioletto, Andrea Fogari, Luca Ghidelli, Kirill Kuzmin, Giovanni Paolini, Pietro Vertechi
Observer A	Massimo Gobbino
Observer B	Federico Poloni

  **IRN**

  **IRL**

  **ISR**

 **ITA**

JPN   Japan

Leader	Yuji Ito
Deputy	Yasuharu Asai
Contestants	Kazuhiro Hosaka, Shiro Imamura, Suguru Ishikawa, Akio Kishikawa, Makoto Soejima, Motoki Takigiku
Observer A	Yuki Ito, Yasuo Morita, Yuta Ohashi, Masaki Watanabe
Observer B	Keiko Tasaki

KAZ   Kazakhstan

Leader	Yerzhan Baisalov
Contestants	Yegor Klochkov, Sanzhar Orazbayev, Nursultan Khadjimuratov, Kanat Satylkhanov, Birzhan Muldagaliyev, Tolebi Sailauov
Observer A	Tildash Bituova

PRK   Democratic People's Republic of Korea

Leader	Yong Chol Ham
Deputy	Yong Ho Kim
Contestants	Un Song Ri, Ho Gon Jon, Jong Chol Kim, Yong Hyon Ri, Jang Su Choe, Hae Chol Son

KOR   Republic of Korea

Leader	Myung-Hwan Kim
Deputy	Byeong Kweon Oh
Contestants	Tae Gu Kang, Young Wook Lyoo, Tae Joo Ahn, Sang Hoon Lee, Sunkyu Lim, Hyun Sub Hwang
Observer A	Yong Jin Song
Observer B	Hyunkyung Ku, Sang-Il Oum, Hee One Park, Jeha Yang

JPN

KAZ

PRK

KOR

KWT   Kuwait

Leader	Ibrahim Alqattan
Deputy	Husain Ali
Contestants	Ali Ahmad Nawab, Ahmad Ameen Alshemali, Hessah Abdullatif Albanwan, Hadeel Awadh Abdullah
Observer A	Fatemah Qasem

KGZ   Kyrgyzstan

Leader	Buras Boljiev
Deputy	Ainagul Osmonalieva
Contestants	Azamat Zhabykeev, Radik Srajidinov, Ilkhomzhon Kalandarov, Azamat Askarov, Susarbek Baibaev, Meerim Topchubaeva

LVA   Latvia

Leader	Agnis Andžāns
Deputy	Juris Škuškovniks
Contestants	Pēteris Eriņš, Ēriks Gopaks, Ieva Ozola, Jānis Smilga, Jevgēnijs Vihrovs, Normunds Vilciņš
Observer A	Laili Sakijeva

LIE   Liechtenstein

Leader	Julian Kellerhals
Deputy	Reto Locher
Contestants	Ricarda Gassner, Florian Meier

 **KWT**

  **KGZ**

  **LVA**

 **LIE**

LTU   Lithuania

Leader	Artūras Dubickas
Deputy	Romualdas Kašuba
Contestants	Vaidotas Juronis, Paulius Kantautas, Edvard Poliakov, Rolandas Glotnis, Greta Kuprijanovaitė, Linas Klimavičius
Observer A	Juozas Juvencijus Macys

LUX   Luxembourg

Leader	Charles Leytem
Deputy	Hsu Ming-Koon
Contestants	Pierre Haas, Jerome Urhausen, Jingran Lin, Philippe Schram, Marc Sinner, Gregoire Genest

MAC   Macau

Leader	Ieng Tak Leong
Deputy	Lung Yam Wan
Contestants	Chi Choi Wong, Hou Meng Ip, Iat Kei Chan, Chi Tou Lam, Chao Keong Lo, Kan Chun Leong

MKD   The former Yugoslav Republic of Macedonia

Leader	Vesna Manova-Erakovik
Deputy	Alekso Malcheski
Contestants	Bodan Arsovski, Filip Talimdzioski, Predrag Gruevski, Bojan Joveski, Stefan Lozanovski, Stefan Stojchevski

  **LTU**

  **LUX**

  **MAC**

MKD

MAS   Malaysia

Leader   M. Suhaimi Ramly
Contestants   Loke Zhi Kin, Muhammad Syafiq Johar

MRT   Mauritania

Leader   Isselmou Ould Lebat Ould Farajou
Deputy   Horma Ould Hamoud
Contestants   Cheikh Tidjani Ahmed Vall, Ahmed Salem Sid'Ahmed, Sidi Mohamed Ahmed
              Salem, Mohamed Salem Abdellahi, Mouadh Ahmed Bah, Saadna Mohamed
              Vadel

MEX   Mexico

Leader   Radmila Bulajich
Deputy   Pablo Soberón
Contestants   Erik Alejandro Gallegos Baños, Manuel Guillermo López Buenfil, César Bibiano
              Velasco, Flavio Hernández González, Luis Ángel Isaías Castellanos, César
              Ernesto Rodríguez Angón
Observer A   Rogelio Valdez

MDA   Republic of Moldova

Leader   Valeriu Baltag
Deputy   Iurie Boreico
Contestants   Dragoş Gîlcă, Iulian Gramaţki, Alexandru Grigoroi, Andrei Iliaşenco, Mihai
              Indricean, Andrei Ivanov

MAS

MRT

MEX

MDA

MNG   Mongolia

Leader   Dashdorj Tserendorj
Contestants   Batkhuyag Batsaikhan,   Enkhzaya Enkhtaivan,   Luvsanbyamba Buyankhuu,
              Tsogbayar Idertsogt, Zorigoo Ochirkhuyag, Byambadorj Otgonsuren
Observer B   Baldorj Sandagdorj

MNE   Montenegro

Leader   Svjetlana Terzić
Deputy   Savo Tomović
Contestants   Sonja Vuksanovic, Milorad Vujkovic, Radovan Krtolica, Nikola Milinkovic

MAR   Morocco

Leader   Abdelmoumen Med El Rhezzali
Deputy   El Maachi Laabiyad
Contestants   Naoufal Aghbal,   Abdelatif Mharchi,   Abdelilah El Hadfaoui,   Rim Hariss,
              Amine Boutaybi, Mohammed Benslimane
Observer A   Nouzha El Yacoubi

NLD   Netherlands

Leader   Quintijn Puite
Deputy   Birgit van Dalen
Contestants   Wouter Berkelmans, Raymond van Bommel, Harm Campmans, Saskia Cham-
              bille, David Kok, Maarten Roelofsma
Observer A   Wim Berkelmans, Tom Verhoeff, Gerhard Woeginger, Hans van Duijn, Ronald van
              Luijk
Observer B   Jelle Loois
Observer C   Merlijn Staps, Theo Wesker, Lidy Wesker-Elzinga

  **MNG**

  **MNE**

  **MAR**

  **NLD**

NZL   New Zealand

Leader	Michael Albert
Deputy	Heather Rose Macbeth
Contestants	Malcolm Granville, Ben Kornfeld, Stephen Mackereth, David Shin, Ha Young Shin, Michael Wang
Observer A	Chris Tuffley
Observer C	May Meng

NGA   Nigeria

Leader	Samson Olatunji Ale
Deputy	Oluwaniyi Steve Dele
Contestants	Isa Modibbo Ismail, Damilola Durajaiye, Ogunkola Opemipo Oladapo, Mark Donald Kanayochukwu Amobi, Jombo Eniweke Eunice, Puis Aje Onah
Observer B	Adewale Roland Tunde Solarin

NOR   Norway

Leader	Dávid Kunszenti-Kovács
Deputy	Pål Hermunn Johansen
Contestants	Sivert Bocianowski, Karl Erik Holter, Sondre Kvamme, Bernt Ivar Nødland, Felix Tadeus Prinz, Jarle Stavnes

PAK   Pakistan

Leader	Alla Ditta Choudary
Deputy	Ahmad Mahmood Qureshi
Contestants	Waqar Ali Syed, Isfar Tariq, Ahmad Bilal Aslam, Absar Ul Hassan, Qasim Mahmood
Observer A	Barbu Berceanu, Valiantsina Mamayeva
Observer B	Ejaz Hussain Malik

NZL

NGA

NOR

PAK

PAN   Panama

Leader   Jaime Gutierrez
Contestants   Antonio Fan

PAR   Paraguay

Leader   Jose Guillermo von Lucken
Deputy   Diana Gimenez de von Lucken
Contestants   Ariel Schvartzman Cohenca,   Marcos Martínez Sugastti,   Juan José Mongelós
              Wollmeister, Claudia Vanessa Montanía Portillo

PER   Peru

Leader   Emilio Gonzaga Ramírez
Deputy   Mariano Adán González Ulloa
Contestants   César Cuenca Lucero, Julián Alonso Mejía Cordero, Percy Augusto Guerra Ríos,
              Ricardo Jesús Ramos Castillo,   Raúl Arturo Chávez Sarmiento,   Tomás Miguel
              Angles Larico
Observer C   Thonchy Cristiam Romero Vilela, Angel Sánchez Oré

PHI   Philippines

Leader   Ian June Luzon Garces
Deputy   Julius Magalona Basilla
Contestants   Carlo Francisco Echavez Adajar,   Earl John Rosales Chua,   Carmela Antoinette
              Sio Lao, Jonathan Santos Wong

PAN

PAR

PER

PHI

POL   Poland

Leader	Andrzej Komisarski
Deputy	Przemysław Mazur
Contestants	Tomasz Kociumaka, Karol Konaszyński, Jakub Oćwieja, Damian Orlef, Tomasz Pawłowski, Jakub Witaszek

POR   Portugal

Leader	Joana Teles
Deputy	Joel Moreira
Contestants	Gonçalo Pereira Simões Matos, João Morais Carreira Pereira, Jorge Ricardo Landeira da Silva Miranda, Pedro Manuel Passos de Sousa Vieira, Raúl Queiroz Do Vale de Noronha Penaguião, Ricardo Correia Moreira

PRI   Puerto Rico

Leader	Luis Caceres
Deputy	Yuri Rojas
Contestants	Aravind Arun, George Arzeno, Kidhanis De Jesus, Jose Pacheco, Sara Rodriguez, Alan Wagner
Observer B	Alfredo Villanueva

ROU   Romania

Leader	Radu Gologan
Deputy	Mihai Băluna
Contestants	Elena Mădălina Persu, Andrei Deneanu, Francisc Bozgan, Tudor Pădurariu, Marius Tiba, Omer Cerrahoglu
Observer A	Dan Schwarz
Observer B	Cristian Alexandrescu

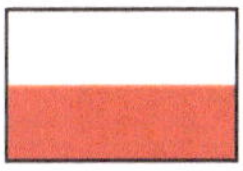

 POL

 POR

PRI

ROU

RUS   Russian Federation

Leader	Nazar Agakhanov
Deputy	Dmitry Tereshin
Contestants	Vladimir Bragin, Marsel Matdinov, Gleb Nenashev, Victor Omelyanenko, Kirill Savenkov, Konstantin Tyshchuk
Observer A	Ilya Bogdanov, Pavel Kozhevnikov
Observer B	Maxim Pratusevich

SLV   El Salvador

Leader	Carlos Mauricio Canjura Linares
Deputy	Riquelmi Salvador Cardona Fuentes
Contestants	Julio Cesar Ayala Menjivar, Jaime Antonio Bermudez Huezo, Nahomy Jhopselyn Hernández Cruz

SRB   Serbia

Leader	Đorđe Krtinić
Deputy	Dušan Đukić
Contestants	Teodor von Burg, Luka Milićević, Dušan Milijančević, Vukašin Stojisavljević, Mihajlo Cekić, Stefan Stojanović

SGP   Singapore

Leader	Yan Loi Wong
Deputy	David Chia
Contestants	Jia-Han Chiam, Tian Wen Daniel Low, Yao Chen Ivan Loh, Jia Hao Barry Tng, Ryan Jun Neng Chan, Jeck Lim
Observer A	Kim Hoo Hang, Tiong Seng Tay
Observer B	Jun Wei Ho, Oi Mei Teo, Teck Kian Teo, Jiawei Wu

RUS

SLV

SRB

SGP

SVK   Slovakia

Leader	Vojtech Bálint
Deputy	Ján Mazák
Contestants	Martin Bachratý, Peter Csiba, Eduard Eiben, Michal Hagara, Filip Sládek, Jakub Uhrík
Observer A	Dagmar Bálintová, Peter Novotný
Observer B	Oliver Ralík

SVN   Slovenia

Leader	Gregor Dolinar
Deputy	Irena Majcen
Contestants	Matej Aleksandrov, Gregor Grasselli, Nik Jazbinšek, Anja Komatar, Matjaž Leonardis, Primož Pušnik

SAF   South Africa

Leader	David Hatton
Deputy	Maciej Stankiewicz
Contestants	Francois Conradie, Henry Thackeray, Liam Baker, Sean Wentzel, Arlton Gilbert, Greg Jackson
Observer A	John Webb, Anthea Webb
Observer B	Phil Labuschagne

ESP   Spain

Leader	María Gaspar
Deputy	Ignasi Mundet Riera
Contestants	Moisés Herradón Cueto, Iván Geffner, Jaime Roquero, Glenier Lázaro Bello Burguet, Ander Lamaison, Alberto Merchante
Observer A	Marco Castrillón

SVK

SVN

SAF

ESP

LKA   Sri Lanka

Leader   Chanakya Janak Wijeratne
Deputy   Thameera Priyadarshi Senanayaka
Contestants   Lajanugan Logeswaran, Buddhima Ruwanmini Gamlath Gamlath Ralalage, Mahamarakkalage Dileepa Yasas Fernando, Kirshanthan Sundararajah, Don Anton Tharindu Kumar Warnakula Warnakulaarachchiralalage, Mukundadura Yasod Sankalpa Fonseka

SWE   Sweden

Leader   Paul Vaderlind
Deputy   Victor Ufnarovski
Contestants   Hampus Engsner, Gabriel Isheden, Jenny Johansson, Eric Larsson, Rickard Norlander, Peter Zarén

SUI   Switzerland

Leader   Anna Devic
Deputy   Thomas Huber
Contestants   Jürg Bachmann, Hrvoje Dujmovic, Eben Freeman, Clemens Pohle, Raphael Steiner, Pascal Su

SYR   Syria

Leader   Abdullatif Hanano
Deputy   Salah Asad
Contestants   Budour Khalil, Basel Ahmad, Nour Maamary, Mokhtar Nadren, Dina Ibrahim

LKA

SWE

SUI

SYR

TWN   Taiwan

Leader	Shu-Chung Liu
Deputy	Shou-Jen Hsiao
Contestants	Han Lin Hsieh, Yu-Fan Tung, Chien-Yi Wang, Hsin-Po Wang, Yi-Chan Wu, Chen-Yu Yang
Observer A	Cheng-Der Fuh
Observer B	Yeong-Nan Yeh
Observer C	Cheng-Chiang Tsai

TJK   Tajikistan

Leader	Umed Karimov
Deputy	Yusuf Balkash
Contestants	Olimjon Pirahmad, Muhamadjon Shoev, Inomzhon Mirzaev, Komron Giesiev, Shohin Sadullozoda, Khusravi Elmurod

THA   Thailand

Leader	Paisan Nakmahachalasint
Deputy	Nattaphan Kittisin
Contestants	Nipun Pitimanaaree, Pakawut Jiradilok, Pongpak Bhumiwat, Supanat Kamtue, Suthee Ruangwises, Thanard Kurutach
Observer A	Yotsanan Meemark, Jittichai Rudjanakanoknad
Observer B	Danita Chunarom

TTO   Trinidad and Tobago

Leader	Indra Haraksingh
Deputy	Christopher Brereton
Contestants	Chandresh Amrit Ramlagan, Kerry Shastri Singh, Prithvi Ramakrishnan, Bjorn Varuun Ramroop, Amanda, Mary Aleong, Siddhartha Jahorie

TWN

TJK

THA

TTO

TUN   Tunisia

Leader   Seifeddine Snoussi
Contestants   Beyrem Khalfaoui, Amine Marrakchi, Imen Allouch, Mannai Nidhal, Youssef
              Khalfalli

TUR   Turkey

Leader   Okan Tekman
Deputy   Azer Kerimov
Contestants   Vefa Göksel, Fehmi Emre Kadan, Ufuk Kanat, Süreyya Emre Kurt, Melih
              Üçer, Umut Varolgüneş

TKM   Turkmenistan

Leader   Erol Aslan
Deputy   Jumageldi Chariyev
Contestants   Nazar Emirov, Kakamyrat Tushiyev, Gaygysyz Hojanazarov, Dovlet Akyniya-
              zov, Archyn Seyidov, Agageldi Samedov

UKR   Ukraine

Leader   Bogdan Rublov
Deputy   Nataliia Prokopenko
Contestants   Igor Kudla, Anastasiya Lysakevych, Vitaliy Senin, Nazar Serdyuk, Darya
              Shchedrina, Bogdan Veklych
Observer A   Anton Mellit, Sergiy Torba

TUN

TUR

TKM

UKR

UAE   United Arab Emirates

Leader   Mohamed Salim Almakhooli
Deputy   Aaesha Jasem Ali Alshamsi
Contestants Alya Alqaydi, Fatmah Al Salami, Kaltham Yousef Tahnon, Meera Saeed Khalifa Almehairi, Alya Abdulla Musabbeh Salem Alqaydi

UNK   United Kingdom

Leader   Geoff Smith
Deputy   Vesna Kadelburg
Contestants Chris Bellin, Luke Betts, Tim Hennock, Peter Leach, Sean Moss, Preeyan Parmar
Observer A   James Cranch
Observer B   Mary Wimbury
Observer C   Jacqui Lewis

USA   United States of America

Leader   Zuming Feng
Deputy   Josh Nichols-Barrar
Contestants John Berman, Wenyu Cao, Eric Larson, Delong Meng, Evan O'Dorney, Qinxuan Pan
Observer B   Steven Dunbar

URY   Uruguay

Leader   Leonardo Lois
Deputy   Eduardo Peraza
Contestants Germán Chiazzo, Matias Escuder, Ari Najman, Javier Peraza, Ismael Valentín Rodríguez Brena, Nicolas Uviedo

  **UAE**

  **UNK**

  **USA**

  **URY**

UZB   Uzbekistan

Leader   Shuhrat Ismailov
Contestants Abror Pirnapasov,   Azizkhon Nazarov,   Diyora Salimova,   Gulomjon Ab-
             durashitov, Ibrokhimbek Akramov, Jasurbek Bahramov

VEN   Venezuela

Leader   Rafael Sánchez Lamoneda
Deputy   Laura Vielma Herrero
Contestants Carmela Acevedo, Mauricio Marcano

VNM   Vietnam

Leader   Hà Huy Khoái
Deputy   Nguyễn Khắc Minh
Contestants Nguyễn Xuân Cương,   Hà Khương Duy,   Nguyễn Hoàng Hải,   Phạm Hy Hiếu,
             Phạm Đức Hùng,   Tạ Đức Thành
Observer C   Đào Mạnh Thắng, Hạ Vũ Anh, Tô Xuân Hải, Lê Đức Thịnh

ZWE   Zimbabwe

Leader   Thomas Masiwa
Deputy   John Walter Greenacre
Contestants Mahdi Finnigan, Oscar Takabvirwa

UZB

VEN

VNM

ZWE

5.2 Observer Countries

Ivory Coast   **CIV**

Observer A   Isaac Tape

Saudi Arabia   **SAU**

Observer A   Abdulaziz Al-Harthi, Fawzi Al-Thukair
Observer B   Mahmoud Nagadi

Senegal   **SEN**

Observer A   Samba Dabo

Chapter 6
Mathematical Competition

According to the regulations, every participating country of an IMO is invited to send suggestions for the competition problems. For the 50th IMO 2009 the German problem selection committee (PSC) received 132 problems from 39 countries. The 12-member PSC, under the leadership of Jürgen Prestin who was also Chief Coordinator and Vice Chairman of the Jury, worked on the problems for about 3 months, gradually reducing the total number of submitted problems to a minimum of 30 problems for the shortlist, from which the Jury had to select 6 problems. The PSC chose 8 problems from the fields of both Combinatorics and Geometry, and 7 problems each on Algebra and Number Theory. In addition the PSC tried to provide all possible ways of solving these problems. The shortlist contained up to 5 different solutions. In some cases the formulations of the question-setters were improved or alternative versions were added. The now renowned grasshopper problem (Problem 6) submitted by Russia was generalized by Christian Reiher, and the Jury chose this revised version. The 6 Problem Captains (all members of the PSC) were: Bernd Mulansky (Problem 1), Martin Welk (Problem 2), Andreas Felgenhauer (Problem 3), Karl Fegert (Problem 4), Robert Strich (Problem 5) and Konrad Engel (Problem 6). Several new solutions were specially created by Christian Reiher and Peter Scholze. Eckart Specht prepared all the geometrical figures, and Roger Labahn was responsible for the LATEX layout of the shortlist booklet. The complete shortlist will be available at http://www.imo-official.org when the 51st IMO starts in July 2010. The Jury excluded one problem from the shortlist, as it emerged that a similar problem had already been used in a recent Olympiad. The Jury likes to include one easy, one medium and one hard problem on each day of the competition. The outcome of the 50th IMO in 2009 is shown in the table below. Problem 6 was the second hardest in the history of the IMO with only 2.4% of the possible points awarded for this problem. Only problem 6 of the 48th IMO 2007 in Vietnam was a little bit harder with 2.2% of all possible points being awarded.

Problem	Field	Result
1	Number Theory	68.6%
2	Geometry	53.0%
3	Algebra	14.6%
4	Geometry	41.6%
5	Algebra	35.3%
6	Combinatorics	2.4%
total		35.9%

   After selecting the six problems the Jury discussed the final wording and prepared the translations into the official languages. These five versions are presented in the next Section 6.1. All the other 50 language versions were prepared by the appropriate team leaders and then approved by the Jury. All translations are now available at `http://www.imo-official.org`. Section 6.2 gives the solutions to the six IMO problems. In some cases the Problem Captains added comments or rewrote the solutions based on the solutions suggested by the participants. Section 6.3 lists all award winners and Section 6.4 all individual scores.

6.1 Problems

 Language: **English**

Day: 1

Wednesday, July 15, 2009

Problem 1. Let n be a positive integer and let $a_1, \ldots, a_k$ ($k \geq 2$) be distinct integers in the set $\{1, \ldots, n\}$ such that n divides $a_i(a_{i+1} - 1)$ for $i = 1, \ldots, k-1$. Prove that n does not divide $a_k(a_1 - 1)$.

Problem 2. Let ABC be a triangle with circumcentre O. The points P and Q are interior points of the sides CA and AB, respectively. Let K, L and M be the midpoints of the segments BP, CQ and PQ, respectively, and let Γ be the circle passing through K, L and M. Suppose that the line PQ is tangent to the circle Γ. Prove that $OP = OQ$.

Problem 3. Suppose that $s_1, s_2, s_3, \ldots$ is a strictly increasing sequence of positive integers such that the subsequences

$$s_{s_1}, s_{s_2}, s_{s_3}, \ldots \qquad \text{and} \qquad s_{s_1+1}, s_{s_2+1}, s_{s_3+1}, \ldots$$

are both arithmetic progressions. Prove that the sequence $s_1, s_2, s_3, \ldots$ is itself an arithmetic progression.

Thursday, July 16, 2009

Problem 4. Let ABC be a triangle with $AB = AC$. The angle bisectors of $\angle CAB$ and $\angle ABC$ meet the sides BC and CA at D and E, respectively. Let K be the incentre of triangle ADC. Suppose that $\angle BEK = 45°$. Find all possible values of $\angle CAB$.

Problem 5. Determine all functions f from the set of positive integers to the set of positive integers such that, for all positive integers a and b, there exists a non-degenerate triangle with sides of lengths

$$a, \ f(b) \ \text{and} \ f(b + f(a) - 1).$$

(A triangle is *non-degenerate* if its vertices are not collinear.)

Problem 6. Let $a_1, a_2, \ldots, a_n$ be distinct positive integers and let M be a set of $n - 1$ positive integers not containing $s = a_1 + a_2 + \cdots + a_n$. A grasshopper is to jump along the real axis, starting at the point 0 and making n jumps to the right with lengths $a_1, a_2, \ldots, a_n$ in some order. Prove that the order can be chosen in such a way that the grasshopper never lands on any point in M.

Mercredi 15 juillet 2009

Problème 1. Soit n un entier strictement positif et soit $a_1, \ldots, a_k$, avec $k \geqslant 2$, des entiers strictement positifs distincts appartenant à l'ensemble $\{1, \ldots, n\}$ tels que n divise $a_i(a_{i+1} - 1)$ pour $i = 1, \ldots, k-1$.

Montrer que n ne divise pas $a_k(a_1 - 1)$.

Problème 2. Soit ABC un triangle et O le centre de son cercle circonscrit. Les points P et Q sont des points intérieurs aux côtés CA et AB respectivement. Soit K, L et M les milieux respectifs des segments BP, CQ et PQ, et soit Γ le cercle passant par K, L et M. On suppose que la droite (PQ) est tangente au cercle Γ.

Montrer que $OP = OQ$.

Problème 3. Soit $s_1, s_2, s_3, \ldots$ une suite strictement croissante d'entiers strictement positifs telle que les sous-suites

$$s_{s_1}, s_{s_2}, s_{s_3}, \ldots \qquad \text{et} \qquad s_{s_1+1}, s_{s_2+1}, s_{s_3+1}, \ldots$$

soient deux progressions arithmétiques.

Montrer que la suite $s_1, s_2, s_3, \ldots$ est aussi une progression arithmétique.

Jeudi 16 juillet 2009

Problème 4. Soit ABC un triangle tel que $AB = AC$. Les bissectrices de $\widehat{CAB}$ et $\widehat{ABC}$ rencontrent respectivement les côtés BC et CA en D et E. Soit K le centre du cercle inscrit dans le triangle ADC. On suppose que $\widehat{BEK} = 45°$.

Trouver toutes les valeurs possibles de $\widehat{CAB}$.

Problème 5. Déterminer toutes les fonctions f de l'ensemble des entiers strictement positifs dans l'ensemble des entiers strictement positifs telles que, pour tous entiers strictement positifs a et b, il existe un triangle non aplati dont les longueurs des côtés sont

$$a, \ f(b) \ \text{et} \ f(b + f(a) - 1).$$

Problème 6. Soit $a_1, a_2, \ldots, a_n$ des entiers strictement positifs distincts et soit M un ensemble de $n - 1$ entiers strictement positifs ne contenant pas $s = a_1 + a_2 + \cdots + a_n$. Une sauterelle doit faire des sauts le long de l'axe réel ; partant du point 0, elle doit effectuer n sauts vers la droite de longueurs $a_1, a_2, \ldots, a_n$ dans l'ordre de son choix.

Montrer que la sauterelle peut choisir l'ordre de ses sauts de façon à ne passer par aucun point de M.

Mittwoch, 15. Juli 2009

Aufgabe 1. Es seien n und k positive ganze Zahlen mit $k \geq 2$. Ferner seien $a_1, \ldots, a_k$ paarweise verschiedene ganze Zahlen aus der Menge $\{1, \ldots, n\}$ derart, dass n die Zahl $a_i(a_{i+1} - 1)$ für jedes $i = 1, \ldots, k - 1$ teilt. Man zeige, dass dann n die Zahl $a_k(a_1 - 1)$ nicht teilt.

Aufgabe 2. Es sei ABC ein Dreieck mit Umkreismittelpunkt O. Es seien P und Q innere Punkte der Seiten CA und AB. Ferner seien K, L und M die Mittelpunkte der Strecken BP, CQ bzw. PQ. Der Kreis Γ gehe durch K, L und M. Die Gerade PQ sei Tangente an den Kreis Γ. Man zeige, dass $|OP| = |OQ|$ gilt.

Aufgabe 3. Es sei $s_1, s_2, s_3, \ldots$ eine streng monoton wachsende Folge positiver ganzer Zahlen derart, dass die beiden Teilfolgen

$$s_{s_1}, \, s_{s_2}, \, s_{s_3}, \, \ldots \quad \text{und} \quad s_{s_1+1}, \, s_{s_2+1}, \, s_{s_3+1}, \, \ldots$$

jeweils arithmetische Folgen sind. Man zeige, dass $s_1, s_2, s_3, \ldots$ ebenfalls eine arithmetische Folge ist.

Language: German   *Zeit: $4\frac{1}{2}$ Stunden*
                                Bei jeder Aufgabe sind 7 Punkte erreichbar.

Donnerstag, 16. Juli 2009

Aufgabe 4. Es sei ABC ein Dreieck mit $|AB| = |AC|$. Die Innenwinkelhalbierenden der Winkel BAC und CBA schneiden die Seiten BC und AC in den Punkten D bzw. E. Es sei K der Inkreismittelpunkt des Dreiecks ADC. Ferner sei $\sphericalangle BEK = 45°$. Man bestimme alle möglichen Werte von $\sphericalangle BAC$.

Aufgabe 5. Man bestimme alle Funktionen f, die auf der Menge der positiven ganzen Zahlen definiert sind und nur positive ganze Zahlen als Werte annehmen, so dass es für alle positiven ganzen Zahlen a und b ein nicht entartetes Dreieck mit Seitenlängen

$$a, \; f(b) \text{ und } f(b + f(a) - 1)$$

gibt.
(Ein Dreieck heißt *nicht entartet*, wenn seine Eckpunkte nicht kollinear sind.)

Aufgabe 6. Es seien n eine positive ganze Zahl, $a_1, a_2, \ldots, a_n$ paarweise verschiedene positive ganze Zahlen und M eine Menge von $n-1$ positiven ganzen Zahlen, die nicht die Summe $s = a_1 + a_2 + \ldots + a_n$ als Element enthält. Ein Grashüpfer springt längs der reellen Zahlengerade. Er startet im Nullpunkt und vollführt n Sprünge nach rechts mit Längen $a_1, a_2, \ldots, a_n$ in beliebiger Reihenfolge. Man zeige, dass der Grashüpfer seine Sprünge so anordnen kann, dass er nie auf einem Punkt aus M landet.

Language: **Russian**

Day: 1

Среда, 15 июля 2009 г.

Задача 1. Даны целое положительное число n и попарно различные целые числа $a_1, \ldots, a_k$ ($k \geq 2$) из множества $\{1, \ldots, n\}$ такие, что для каждого $i = 1, \ldots, k-1$ число $a_i(a_{i+1} - 1)$ делится на n. Докажите, что число $a_k(a_1 - 1)$ не делится на n.

Задача 2. Точка O — центр окружности, описанной около треугольника ABC. Пусть P и Q — внутренние точки отрезков CA и AB соответственно. Точки K, L и M — середины отрезков BP, CQ и PQ соответственно, а Γ — окружность, проходящая через точки K, L и M. Известно, что прямая PQ касается окружности Γ. Докажите, что $OP = OQ$.

Задача 3. Дана строго возрастающая последовательность целых положительных чисел $s_1, s_2, s_3, \ldots$ такая, что каждая из двух последовательностей

$$s_{s_1}, s_{s_2}, s_{s_3}, \ldots \qquad \text{и} \qquad s_{s_1+1}, s_{s_2+1}, s_{s_3+1}, \ldots$$

является арифметической прогрессией. Докажите, что последовательность $s_1, s_2, s_3, \ldots$ также является арифметической прогрессией.

Четверг, 16 июля 2009 г.

Задача 4. Треугольник ABC таков, что $AB = AC$. Биссектрисы углов CAB и ABC пересекают стороны BC и CA в точках D и E соответственно. Обозначим через K центр окружности, вписанной в треугольник ADC. Оказалось, что $\angle BEK = 45°$. Найдите все возможные значения угла CAB.

Задача 5. Найдите все функции $f: \mathbb{N} \to \mathbb{N}$ (то есть функции, определенные на множестве всех целых положительных чисел и принимающие целые положительные значения) такие, что для любых целых положительных a и b существует невырожденный треугольник, длины сторон которого равны трем числам

$$a, \quad f(b) \quad \text{и} \quad f(b + f(a) - 1).$$

(Треугольник называется *невырожденным*, если его вершины не лежат на одной прямой.)

Задача 6. Даны попарно различные целые положительные числа $a_1, a_2, \ldots, a_n$, а также множество M, состоящее из $n - 1$ целого положительного числа, но не содержащее число $s = a_1 + a_2 + \ldots + a_n$. Кузнечик должен сделать n прыжков вправо по числовой прямой, стартуя из точки с координатой 0. При этом длины его прыжков должны равняться числам $a_1, a_2, \ldots, a_n$, взятым в некотором порядке. Докажите, что этот порядок можно выбрать таким образом, чтобы кузнечик ни разу не приземлился в точке, имеющей координату из множества M.

th International
Mathematical
Olympiad

Language: **Spanish**

Day: 1

Miércoles 15 de julio de 2009

Problema 1. Sea n un entero positivo y sean $a_1, \ldots, a_k$ $(k \geq 2)$ enteros distintos del conjunto $\{1, \ldots, n\}$, tales que n divide a $a_i(a_{i+1} - 1)$, para $i = 1, \ldots, k - 1$. Demostrar que n no divide a $a_k(a_1 - 1)$.

Problema 2. Sea ABC un triángulo con circuncentro O. Sean P y Q puntos interiores de los lados CA y AB, respectivamente. Sean K, L y M los puntos medios de los segmentos BP, CQ y PQ, respectivamente, y Γ la circunferencia que pasa por K, L y M. Se sabe que la recta PQ es tangente a la circunferencia Γ. Demostrar que $OP = OQ$.

Problema 3. Sea $s_1, s_2, s_3, \ldots$ una sucesión estrictamente creciente de enteros positivos tal que las subsucesiones

$$s_{s_1}, s_{s_2}, s_{s_3}, \ldots \qquad \text{y} \qquad s_{s_1+1}, s_{s_2+1}, s_{s_3+1}, \ldots$$

son ambas progresiones aritméticas. Demostrar que la sucesión $s_1, s_2, s_3, \ldots$ es también una progresión aritmética.

Language: Spanish   *Tiempo: 4 horas y 30 minutos*
                                                 Cada problema vale 7 puntos

Jueves 16 de julio de 2009

Problema 4.   Sea ABC un triángulo con $AB = AC$. Las bisectrices de los ángulos $\angle CAB$ y $\angle ABC$ cortan a los lados BC y CA en D y E, respectivamente. Sea K el incentro del triángulo ADC. Supongamos que el ángulo $\angle BEK = 45°$. Determinar todos los posibles valores de $\angle CAB$.

Problema 5.   Determinar todas las funciones f del conjunto de los enteros positivos en el conjunto de los enteros positivos tales que, para todos los enteros positivos a y b, existe un triángulo no degenerado cuyos lados miden

$$a,\ f(b)\ \text{y}\ f(b + f(a) - 1).$$

(Un triángulo es *no degenerado* si sus vértices no están alineados).

Problema 6.   Sean $a_1, a_2, \ldots, a_n$ enteros positivos distintos y M un conjunto de $n - 1$ enteros positivos que no contiene al número $s = a_1 + a_2 + \cdots + a_n$. Un saltamontes se dispone a saltar a lo largo de la recta real. Empieza en el punto 0 y da n saltos hacia la derecha de longitudes $a_1, a_2, \ldots, a_n$, en algún orden. Demostrar que el saltamontes puede organizar los saltos de manera que nunca caiga en un punto de M.

6.2 Solutions

6.2.1 Problem 1 – Number Theory – Australia

Solution 1.

By the assumptions of the problem we have that

$$a_i a_{i+1} \equiv a_i \bmod n \quad \text{for } i = 1, \ldots, k-1,$$

which successively implies

$$a_1 \ldots a_{k-1} a_k \equiv a_1 \ldots a_{k-1} \equiv \cdots \equiv a_1 \bmod n. \tag{6.1}$$

To proceed indirectly, suppose that also $a_k a_1 \equiv a_k \bmod n$. Then we obtain

$$a_1 \ldots a_{k-1} a_k \equiv a_k a_1 \ldots a_{k-1} \equiv a_k a_1 \ldots a_{k-2} \equiv \cdots \equiv a_k a_1 \equiv a_k \bmod n.$$

Therefore, we get $a_1 \equiv a_k \bmod n$. Since $a_1, a_k \in \{1, 2, \ldots, n\}$, this implies $a_1 = a_k$, a contradiction.

Variation 1. Similarly to the derivation of (6.1) we obtain

$$a_2 \ldots a_{k-1} a_k \equiv a_2 \bmod n,$$

which, using $a_k a_1 \equiv a_k \bmod n$ and (6.1), results in $a_1 \equiv a_2 \bmod n$.

Variation 2. Assuming $a_k a_1 \equiv a_k \bmod n$, analogously to (1) we also obtain

$$a_2 \ldots a_k a_1 \equiv a_2 \bmod n$$

$$\vdots$$

$$a_k a_1 \ldots a_{k-1} \equiv a_k \bmod n,$$

hence $a_1 \equiv a_2 \equiv \cdots \equiv a_k \bmod n$. This results in the contradiction $a_1 = \cdots = a_k$.

Solution 2.

Let $k = 2$. From $a_1 \neq a_2$ and $a_1, a_2 \in \{1, 2, \ldots, n\}$ we know that n does not divide $a_1 - a_2$. Thus we have

$$n \nmid a_1 - a_2 + a_1(a_2 - 1) = a_2(a_1 - 1),$$

which is the assertion.

Let $k \geq 3$. From

$$n \mid a_{k-2}(a_{k-1} - 1) \quad \text{and} \quad n \mid a_{k-1}(a_k - 1)$$

it follows that n also divides

$$(1 - a_k) a_{k-2}(a_{k-1} - 1) + a_{k-2} a_{k-1}(a_k - 1) = a_{k-2}(a_k - 1).$$

Continuing in this manner, we get that n divides $a_i(a_k - 1)$ for $i = k - 2, \ldots, 2, 1$. In particular, n divides $a_1(a_k - 1)$. Similarly to the case $k = 2$, this implies $a_1 = a_k$, a contradiction.

Solution 3.

Assume, by contradiction, that n also divides $a_k(a_1 - 1)$.

Let $a_{k+1} := a_1$ and $a_0 := a_k$. Let p^e be one of the prime powers in the decomposition of n, i.e., $p^e \mid n$ and $p^{e+1} \nmid n$. We have

$$p^e \mid a_i(a_{i+1} - 1) \text{ for } i = 1, \ldots, k. \tag{6.2}$$

Claim 1. Either $p^e \mid a_i$ for all i or $p^e \mid a_i - 1$ for all i.
Proof: We have for any i

$$p \nmid a_i \Rightarrow p \mid a_{i+1} - 1 \Rightarrow p \nmid a_{i+1}, \tag{6.3}$$
$$p \mid a_i \Rightarrow p \nmid a_i - 1 \Rightarrow p \mid a_{i-1}. \tag{6.4}$$

Hence in case (6.3) we have $p \nmid a_i$ for all i, so $p^e \mid a_i - 1$ by (6.2).
In case (6.4) we get $p \nmid a_i - 1$ for all i, so $p^e \mid a_i$ by (6.2).

Claim 2. $a_1 \equiv a_2 \equiv \cdots \equiv a_k \bmod n$.
Proof: This follows from claim 1 and the Chinese Remainder Theorem.
Now $a_j \in \{1, 2, \ldots, n\}$ implies that $a_1 = \cdots = a_k$, a contradiction.

Variation 1 for claim 1. Use

$$p \nmid a_i \Rightarrow p^e \mid a_{i+1} - 1 \Rightarrow p \nmid a_{i+1},$$
$$p \nmid a_i - 1 \Rightarrow p^e \mid a_{i-1} \Rightarrow p \nmid a_{i-1} - 1,$$

from which the claim follows immediately.

Variation 2 for claim 1. Replace (6.4) by the following argument:
From (6.3) it follows that it cannot happen that some a_i are divisible by p and some are not. Hence either no a_i is divisible by p or all a_i are divisible by p. In the latter case $p \nmid a_i - 1$.

Variation 3 for claim 1. We have for any i

$$p^e \nmid a_i \Rightarrow p \mid a_{i+1} - 1 \Rightarrow p^e \nmid a_{i+1} \tag{6.5}$$

Either $p^e \mid a_j$ for all j or $p^e \nmid a_j$ for some j. In the latter case we have $p \mid a_i - 1$ for all i by (6.5). Hence $p \nmid a_i$ and therefore $p^e \mid a_i - 1$ for all i.

Variation for claim 2. We have $p^e \mid a_1 - a_2$ for any prime power in the decomposition of n. Hence $n \mid a_1 - a_2$.

Solution 4.

We assume that n divides $a_k(a_1 - 1)$. Then there exist integers $u, v \geq 1$ such that u divides a_k, v divides $a_1 - 1$ and $n = uv$. This implies $\gcd(u, a_k - 1) = 1$ and $\gcd(v, a_1) = 1$. As n, hence u, divides $a_{k-1}(a_k - 1)$, we see that u divides a_{k-1} and $\gcd(u, a_{k-1} - 1) = 1$. Analogously, since v divides $a_1(a_2 - 1)$, it divides $a_2 - 1$ and $\gcd(v, a_2) = 1$.

Inductively, this yields that u divides a_i and v divides $a_i - 1$ for all $i = 1, \ldots, k$. In particular u and v are relatively prime. Taking differences, we get that u and v both divide $a_1 - a_2$. Since $n = uv$ and u and v are relatively prime, n divides $a_1 - a_2$. Now $a_1, a_2 \in \{1, 2, \ldots, n\}$ implies that $a_1 = a_2$, a contradiction.

Solution 5.

Consider the graph with vertices $v_1, \ldots, v_n$ and directed edges from v_i to v_j if and only if $n \mid i(j-1)$. We have to show that this graph does not contain any directed cycle.

Let $n = l$ be the smallest value of n for which the corresponding graph has a directed cycle. First we show that l is a prime power.

If l is not a prime power, it can be written as a product $l = de$ of relatively prime integers greater than 1. Reducing all the numbers modulo d yields a single vertex or a cycle in the corresponding graph on d vertices. Indeed, if $a_i(a_{i+1} - 1) \equiv 0 \mod l$ then this equation also holds modulo d. But since the graph on d vertices has no cycles, by the minimality of l, we must have that all the indices of the cycle are congruent modulo d. The same holds modulo e and hence also modulo $l = de$. But then all the indices are equal, which is a contradiction.

Thus let $l = p^m$ be a prime power. There are no edges ending at v_l, so v_l is not contained in any cycle. All edges not starting at v_l end at a vertex belonging to a non-multiple of p. All edges starting at a non-multiple of p must end at v_1. But there is no edge starting at v_1. Hence there is no cycle.

Problem Captain's Comments on Problem 1.

The proposed formulation of this problem, as submitted by Australia and included in the shortlist, was as follows.

A social club has n members. They have the membership numbers $1, 2, \ldots, n$, respectively. From time to time members send presents to other members, including items they have already received as presents from other members. In order to avoid the embarrassing situation that a member might receive a present that he or she has sent to other members, the club adds the following rule to its statutes at one of its annual general meetings:

"A member with membership number a is permitted to send a present to a member with membership number b if and only if $a(b-1)$ is a multiple of n."

Prove that, if each member follows this rule, none will receive a present from another member that he or she has already sent to other members.

After electing this proposal the first problem of the contest, the jury got into a quite controversial discussion on its wording. Finally, it was decided to turn the "story" down and to prefer a pure mathematical formulation. Your mileage may vary.

The original proposal was used in the online version of the german newspaper "Spiegel" to popularize the IMO 2009. Under the humorous headlines "Formula against embarassing presents" and "Get rid of ugly items", CHRISTIAN REIHER explains the problem (even in a video), see http://www.spiegel.de/wissenschaft/mensch/0, 1518,636670,00.html and http://www.spiegel.de/wissenschaft/mensch/ 0,1518,636682,00.html

Our team of coordinators adapted the shortlist solutions to the changed formulation of the problem. The modified solutions were distributed among all team leaders, and are reproduced above. In the coordination process, it was occasionally delicate to agree whether the contestant covered the more subtle details in Solution 3 or Solution 4.

6.2.2 Problem 2 – Geometry – Russia

Solution 1.

Let K, L, M, B', C' be the midpoints of BP, CQ, PQ, CA, and AB, respectively (see Figure 1). Since $CA \parallel LM$, we have $\angle LMP = \angle QPA$. Since k touches the segment PQ at M, we find $\angle LMP = \angle LKM$. Thus $\angle QPA = \angle LKM$. Similarly it follows from $AB \parallel MK$ that $\angle PQA = \angle KLM$. Therefore, triangles APQ and MKL are similar, hence

$$\frac{AP}{AQ} = \frac{MK}{ML} = \frac{\frac{QB}{2}}{\frac{PC}{2}} = \frac{QB}{PC}. \tag{6.6}$$

Now (1) is equivalent to $AP \cdot PC = AQ \cdot QB$ which means that the power of points P and Q with respect to the circumcircle of $\triangle ABC$ are equal, hence $OP = OQ$.

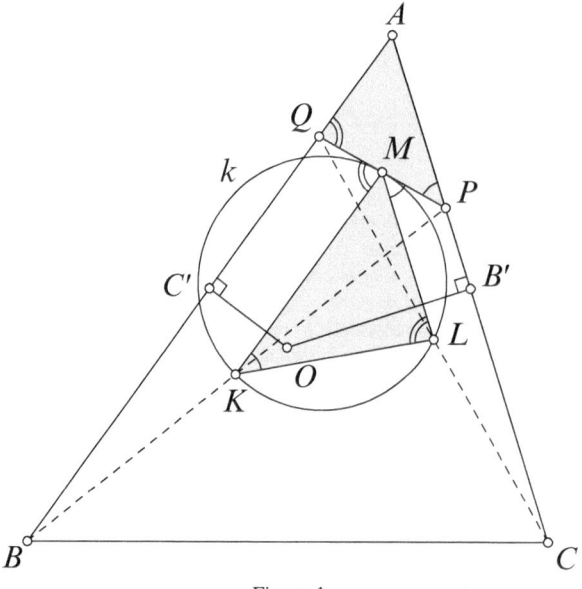

Figure 1

Comment.

The last argument can also be established by the following calculation:

$$\begin{aligned}
OP^2 - OQ^2 &= OB'^2 + B'P^2 - OC'^2 - C'Q^2 \\
&= (OA^2 - AB'^2) + B'P^2 - (OA^2 - AC'^2) - C'Q^2 \\
&= (AC'^2 - C'Q^2) - (AB'^2 - B'P^2) \\
&= (AC' - C'Q)(AC' + C'Q) - (AB' - B'P)(AB' + B'P) \\
&= AQ \cdot QB - AP \cdot PC.
\end{aligned}$$

With (1), we conclude $OP^2 - OQ^2 = 0$, as desired.

Solution 2.

Again, denote by K, L, M the midpoints of segments BP, CQ, and PQ, respectively. Let O, S, T be the circumcenters of triangles ABC, KLM, and APQ, respectively (see Figure 2). Note that MK and LM are the midlines in triangles BPQ and CPQ, respectively, so $\overrightarrow{MK} = \frac{1}{2}\overrightarrow{QB}$ and $\overrightarrow{ML} = \frac{1}{2}\overrightarrow{PC}$. Denote by $\mathrm{pr}_l(\overrightarrow{v})$ the projection of vector $\overrightarrow{v}$ onto line l. Then $\mathrm{pr}_{AB}(\overrightarrow{OT}) = \mathrm{pr}_{AB}(\overrightarrow{OA} - \overrightarrow{TA}) = \frac{1}{2}\overrightarrow{BA} - \frac{1}{2}\overrightarrow{QA} = \frac{1}{2}\overrightarrow{BQ} = \overrightarrow{KM}$ and $\mathrm{pr}_{AB}(\overrightarrow{SM}) = \mathrm{pr}_{MK}(\overrightarrow{SM}) = \frac{1}{2}\overrightarrow{KM} = \frac{1}{2}\mathrm{pr}_{AB}(\overrightarrow{OT})$. Analogously we get $\mathrm{pr}_{CA}(\overrightarrow{SM}) = \frac{1}{2}\mathrm{pr}_{CA}(\overrightarrow{OT})$. Since AB and CA are not parallel, this implies that $\overrightarrow{SM} = \frac{1}{2}\overrightarrow{OT}$.

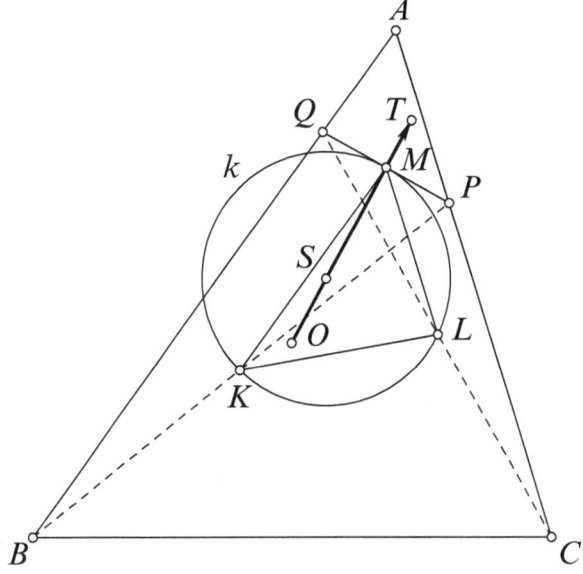

Figure 2

Now, since the circle k touches PQ at M, we get $SM \perp PQ$, hence $OT \perp PQ$. Since T is equidistant from P and Q, the line OT is a perpendicular bisector of segment PQ, and hence O is equidistant from P and Q which finishes the proof.

Problem Captain's Comments on Problem 2

When we started in our team of coordinators to prepare for the marking of Problem 2, the first observation that stroke us was the variety of different approaches to its solution. In addition to the solutions contained in the Shortlist, members of the Jury and coordinators quickly contributed about half a dozen different solutions using the classical instrumentary of elementary geometry, with or without a vectorial formulation.

For example, one could make use of the similarity of triangles KLM and Q_1P_1A, which by the similarity statement of Solution 1 implied that PQQ_1P_1 is a cyclic quadrilateral. Since the perpendicular bisectors of PP_1, QQ_1 are the same as those of AC, AB, respectively, it followed that O is the circumcentre of PQQ_1P_1 an thus the claim. (A caveat of this approach is that one must take care of the case $P = P_1$ or $Q = Q_1$.)

Now it is not uncommon – and certainly a desirable feature – for geometric problems on IMO level to have multiple solutions. Here, the number of basic ideas was not even that large: Virtually every solution had somehow to make use of the parallelity of KM and KL to AB, AC, respectively; and there were essentially two ways to exploit the hypothesis of tangency of k with PQ, namely either via the chord-tangent angle as in Solution 1 or by the orthogonality between tangent and radius, see Solution 2. There were, however, plenty of recombinations of these steps, and coming up with a consistent Marking Scheme that would take into account these possibilities was a challenging task. The resulting "solution graph" evoked some amusement in the Jury but turned out fairly practical in the coordination phase: With few minor deviations, almost all solutions could be represented within it.

Albeit feasible, coordinate approaches to the complete problem were unattractive at large and were rare among the students' solutions, successful ones even rarer.

6.2.3 Problem 3 – Algebra – United States

Solution 1.

Let D be the common difference of the progression $s_{s_1}, s_{s_2}, \ldots$. Let for $n = 1, 2, \ldots$

$$d_n = s_{n+1} - s_n.$$

We have to prove that d_n is constant. First we show that the numbers d_n are bounded. Indeed, by supposition $d_n \geq 1$ for all n. Thus, we have for all n

$$d_n = s_{n+1} - s_n \leq d_{s_n} + d_{s_n+1} + \cdots + d_{s_{n+1}-1} = s_{s_{n+1}} - s_{s_n} = D.$$

The boundedness implies that there exist

$$m = \min\{d_n : n = 1, 2, \ldots\} \quad \text{and} \quad M = \max\{d_n : n = 1, 2, \ldots\}.$$

It suffices to show that $m = M$. Assume that $m < M$. Choose n such that $d_n = m$. Considering a telescoping sum of $m = d_n = s_{n+1} - s_n$ items not greater than M leads to

$$D = s_{s_{n+1}} - s_{s_n} = s_{s_n+m} - s_{s_n} = d_{s_n} + d_{s_n+1} + \cdots + d_{s_n+m-1} \leq mM \qquad (6.7)$$

and equality holds if and only if all items of the sum are equal to M. Now choose n such that $d_n = M$. In the same way, considering a telescoping sum of M items not less than m we obtain

$$D = s_{s_{n+1}} - s_{s_n} = s_{s_n+M} - s_{s_n} = d_{s_n} + d_{s_n+1} + \cdots + d_{s_n+M-1} \geq Mm \qquad (6.8)$$

and equality holds if and only if all items of the sum are equal to m. The inequalities (6.7) and (6.8) imply that $D = Mm$ and that

$$d_{s_n} = d_{s_n+1} = \cdots = d_{s_{n+1}-1} = M \quad \text{if } d_n = m,$$
$$d_{s_n} = d_{s_n+1} = \cdots = d_{s_{n+1}-1} = m \quad \text{if } d_n = M.$$

Hence, $d_n = m$ implies $d_{s_n} = M$. Note that $s_n \geq s_1 + (n-1) \geq n$ for all n and moreover $s_n > n$ if $d_n = n$, because in the case $s_n = n$ we would have $m = d_n = d_{s_n} = M$ in contradiction to the assumption $m < M$. In the same way $d_n = M$ implies $d_{s_n} = m$ and $s_n > n$. Consequently, there is a strictly increasing sequence $n_1, n_2, \ldots$ such that

$$d_{s_{n_1}} = M, \quad d_{s_{n_2}} = m, \quad d_{s_{n_3}} = M, \quad d_{s_{n_4}} = m, \quad \ldots.$$

The sequence $d_{s_1}, d_{s_2}, \ldots$ is the sequence of pairwise differences of $s_{s_1+1}, s_{s_2+1}, \ldots$ and $s_{s_1}, s_{s_2}, \ldots$, hence also an arithmetic progression. Thus $m = M$.

Solution 2.

Let the integers D and E be the common differences of the progressions $s_{s_1}, s_{s_2}, \ldots$ and $s_{s_1+1}, s_{s_2+1}, \ldots$, respectively. Let briefly $A = s_{s_1} - D$ and $B = s_{s_1+1} - E$. Then, for all positive integers n,

$$s_{s_n} = A + nD, \qquad s_{s_n+1} = B + nE.$$

Since the sequence $s_1, s_2, \ldots$ is strictly increasing, we have for all positive integers n

$$s_{s_n} < s_{s_n+1} \leq s_{s_{n+1}},$$

which implies

$$A + nD < B + nE \leq A + (n+1)D,$$

and thereby

$$0 < B - A + n(E - D) \leq D,$$

which implies $D - E = 0$ and thus

$$0 \leq B - A \leq D. \tag{6.9}$$

Let $m = \min\{s_{n+1} - s_n : n = 1, 2, \ldots\}$. Then

$$B - A = (s_{s_1+1} - E) - (s_{s_1} - D) = s_{s_1+1} - s_{s_1} \geq m \tag{6.10}$$

and

$$D = A + (s_1 + 1)D - (A + s_1 D) = s_{s_{s_1}+1} - s_{s_{s_1}} = s_{B+D} - s_{A+D} \geq m(B - A). \tag{6.11}$$

From (6.9) we consider two cases.

Case 1. $B - A = D$.
Then, for each positive integer n, $s_{s_n+1} = B + nD = A + (n+1)D = s_{s_{n+1}}$, hence $s_{n+1} = s_n + 1$ and $s_1, s_2, \ldots$ is an arithmetic progression with common difference 1.

Case 2. $B - A < D$.   Choose some positive integer N such that $s_{N+1} - s_N = m$. Then

$$\begin{aligned}
m(A - B + D - 1) &= m((A + (N+1)D) - (B + ND + 1)) \\
&\leq s_{A+(N+1)D} - s_{B+ND+1} = s_{s_{s_{N+1}}} - s_{s_{s_N+1}+1} \\
&= (A + s_{N+1}D) - (B + (s_N + 1)D) = (s_{N+1} - s_N)D + A - B - D \\
&= mD + A - B - D,
\end{aligned}$$

i.e.,

$$(B - A - m) + (D - m(B - A)) \leq 0. \tag{6.12}$$

The inequalities (6.10)-(6.12) imply that

$$B - A = m \quad \text{and} \quad D = m(B - A).$$

Assume that there is some positive integer n such that $s_{n+1} > s_n + m$. Then

$$m(m+1) \leq m(s_{n+1} - s_n) \leq s_{s_{n+1}} - s_{s_n} = (A + (n+1)D) - (A + nD)) = D = m(B - A) = m^2,$$

a contradiction. Hence $s_1, s_2, \ldots$ is an arithmetic progression with common difference m.

6.2.4 Problem 4 – Geometry – Belgium

Solution 1.

Answer: $\angle BAC = 60°$ or $\angle BAC = 90°$ are possible values and the only possible values.

Let I be the incenter of triangle ABC, then K lies on the line CI. Let F be the point, where the incircle of triangle ABC touches the side AC; then the segments IF and ID have the same length and are perpendicular to AC and BC, respectively.

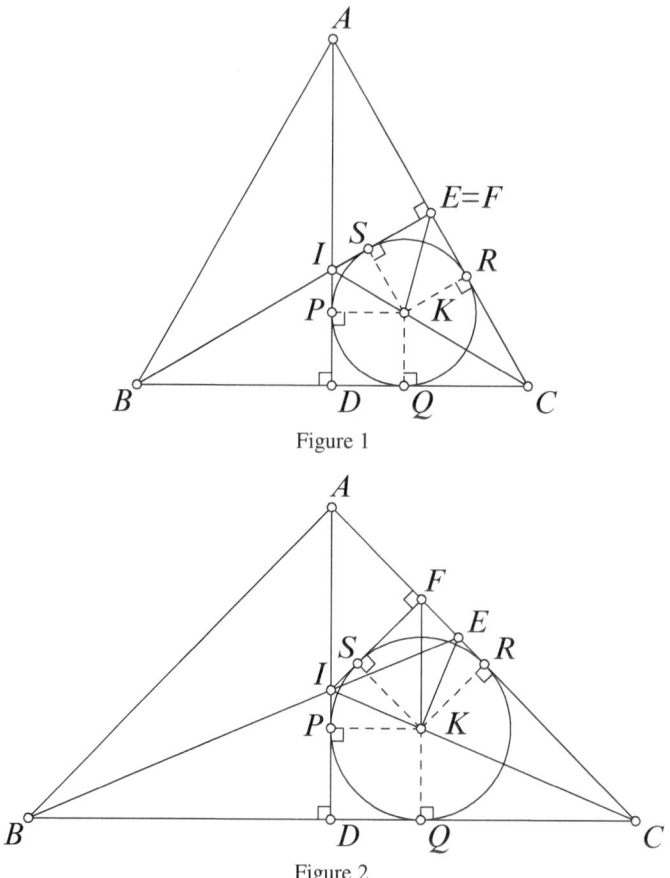

Figure 1

Figure 2

Let P, Q and R be the points where the incircle of triangle ADC touches the sides AD, DC and CA, respectively. Since K and I lie on the angle bisector of $\angle ACD$, the segments ID and IF are symmetric with respect to the line IC. Hence there is a point S on IF where the incircle of triangle ADC touches the segment IF. Then segments KP, KQ, KR and KS all have the same length and are perpendicular to AD, DC, CA and IF, respectively. So – regardless of the value of $\angle BEK$ – the quadrilateral $KRFS$ is a square and $\angle SFK = \angle KFC = 45°$.

Consider the case $\angle BAC = 60°$ (see Figure 1). Then triangle ABC is equilateral. Furthermore we have $F = E$, hence $\angle BEK = \angle IFK = \angle SEK = 45°$. So $60°$ is a possible value for $\angle BAC$.

Now consider the case $\angle BAC = 90°$ (see Figure 2). Then $\angle CBA = \angle ACB = 45°$. Furthermore, $\angle KIE = \frac{1}{2}\angle CBA + \frac{1}{2}\angle ACB = 45°$, $\angle AEB = 180° - 90° - 22.5° = 67.5°$ and $\angle EIA = \angle BID = 180° - 90° - 22.5° = 67.5°$. Hence triangle IEA is isosceles and a reflection of the bisector of

$\angle IAE$ takes I to E and K to itself. So triangle IKE is symmetric with respect to this axis, i.e. $\angle KIE = \angle IEK = \angle BEK = 45°$. So $90°$ is a possible value for $\angle BAC$, too.

If, on the other hand, $\angle BEK = 45°$ then $\angle BEK = \angle IEK = \angle IFK = 45°$. Then

- either $F = E$, which makes the angle bisector BI be an altitude, i.e., which makes triangle ABC isosceles with base AC and hence equilateral and so $\angle BAC = 60°$,
- or E lies between F and C, which makes the points K, E, F and I concyclic, so $45° = \angle KFC = \angle KFE = \angle KIE = \angle CBI + \angle ICB = 2 \cdot \angle ICB = 90° - \frac{1}{2}\angle BAC$, and so $\angle BAC = 90°$,
- or F lies between E and C, then again, K, E, F and I are concyclic, so $45° = \angle KFC = 180° - \angle KFE = \angle KIE$, which yields the same result $\angle BAC = 90°$. (However, for $\angle BAC = 90°$ E lies, in fact, between F and C, see Figure 2. So this case does not occur.)

This proves $90°$ and $60°$ to be the only possible values for $\angle BAC$.

Solution 2.

Denote angles at A, B and C as usual by α, β and γ. Since triangle ABC is isosceles, we have $\beta = \gamma = 90° - \frac{\alpha}{2} < 90°$, so $\angle ECK = 45° - \frac{\alpha}{4} = \angle KCD$. Since K is the incenter of triangle ADC, we have $\angle CDK = \angle KDA = 45°$; furthermore $\angle DIC = 45° + \frac{\alpha}{4}$. Now, if $\angle BEK = 45°$, easy calculations within triangles BCE and KCE yield

$$\angle KEC = 180° - \frac{\beta}{2} - 45° - \beta = 135° - \frac{3}{2}\beta = \frac{3}{2}(90° - \beta) = \frac{3}{4}\alpha,$$
$$\angle IKE = \frac{3}{4}\alpha + 45° - \frac{\alpha}{4} = 45° + \frac{\alpha}{2}.$$

So in triangles ICE, IKE, IDK and IDC we have (see Figure 3)

$$\frac{IC}{IE} = \frac{\sin \angle IEC}{\sin \angle ECI} = \frac{\sin(45° + \frac{3}{4}\alpha)}{\sin(45° - \frac{\alpha}{4})}, \qquad \frac{IE}{IK} = \frac{\sin \angle EKI}{\sin \angle IEK} = \frac{\sin(45° + \frac{\alpha}{2})}{\sin 45°},$$

$$\frac{IK}{ID} = \frac{\sin \angle KDI}{\sin \angle IKD} = \frac{\sin 45°}{\sin(90° - \frac{\alpha}{4})}, \qquad \frac{ID}{IC} = \frac{\sin \angle ICD}{\sin \angle CDI} = \frac{\sin(45° - \frac{\alpha}{4})}{\sin 90°}.$$

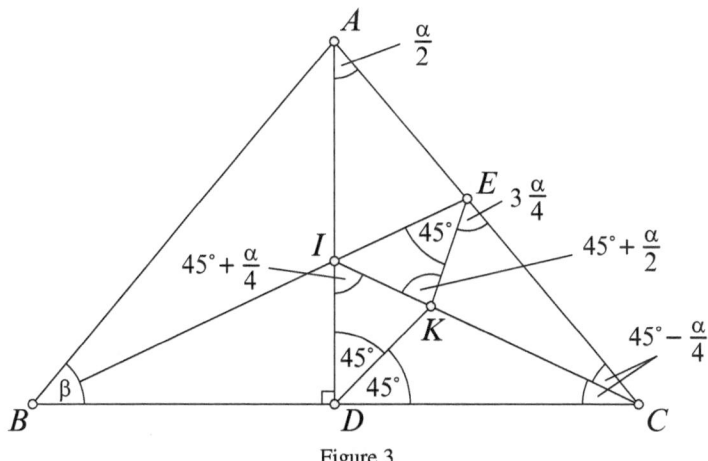

Figure 3

Multiplication of these four equations yields

$$1 = \frac{\sin(45° + \frac{3}{4}\alpha)\sin(45° + \frac{\alpha}{2})}{\sin(90° - \frac{\alpha}{4})}.$$

But, since

$$\sin\left(90° - \tfrac{\alpha}{4}\right) = \cos\tfrac{\alpha}{4} = \cos\left((45° + \tfrac{3}{4}\alpha) - (45° + \tfrac{\alpha}{2})\right)$$
$$= \cos\left(45° + \tfrac{3}{4}\alpha\right)\cos\left(45° + \tfrac{\alpha}{2}\right) + \sin\left(45° + \tfrac{3}{4}\alpha\right)\sin\left(45° + \tfrac{\alpha}{2}\right),$$

this is equivalent to

$$\sin(45° + \tfrac{3}{4}\alpha)\sin(45° + \tfrac{\alpha}{2}) = \cos\left(45° + \tfrac{3}{4}\alpha\right)\cos\left(45° + \tfrac{\alpha}{2}\right) + \sin\left(45° + \tfrac{3}{4}\alpha\right)\sin\left(45° + \tfrac{\alpha}{2}\right)$$

and finally

$$\cos\left(45° + \tfrac{3}{4}\alpha\right)\cos\left(45° + \tfrac{\alpha}{2}\right) = 0.$$

But this means $\cos\left(45° + \tfrac{3}{4}\alpha\right) = 0$, hence $45° + \tfrac{3}{4}\alpha = 90°$, i.e. $\alpha = 60°$ or $\cos\left(45° + \tfrac{\alpha}{2}\right) = 0$, hence $45° + \tfrac{\alpha}{2} = 90°$, i.e. $\alpha = 90°$. So these values are the only two possible values for α.
    On the other hand, both $\alpha = 90°$ and $\alpha = 60°$ yield $\angle BEK = 45°$, this was shown in Solution 1.

6.2.5 Problem 5 – Algebra – France

Solution.

The identity function $f(x) = x$ is the only solution of the problem.
    If $f(x) = x$ for all positive integers x, the given three lengths are x, $y = f(y)$ and $z = f(y + f(x) - 1) = x + y - 1$. Because of $x \geq 1$, $y \geq 1$ we have $z \geq \max\{x,y\} > |x - y|$ and $z < x + y$. From this it follows that a triangle with these side lengths exists and does not degenerate. We prove in several steps that there is no other solution.

    Step 1. We show $f(1) = 1$.
If we had $f(1) = 1 + m > 1$ we would conclude $f(y) = f(y + m)$ for all y considering the triangle with the side lengths 1, $f(y)$ and $f(y + m)$. Thus, f would be m-periodic and, consequently, bounded. Let B be a bound, $f(x) \leq B$. If we choose $x > 2B$ we obtain the contradiction $x > 2B \geq f(y) + f(y + f(x) - 1)$.

    Step 2. For all positive integers z, we have $f(f(z)) = z$.
Setting $x = z$ and $y = 1$ this follows immediately from Step 1.

    Step 3. For all integers $z \geq 1$, we have $f(z) \leq z$.
Let us show, that the contrary leads to a contradiction. Assume $w + 1 = f(z) > z$ for some z. From Step 1 we know that $w \geq z \geq 2$. Let $M = \max\{f(1), f(2), \ldots, f(w)\}$ be the largest value of f for the first w integers. First we show, that no positive integer t exists with

$$f(t) > \frac{z - 1}{w} \cdot t + M, \tag{1}$$

otherwise we decompose the smallest value t as $t = wr + s$ where r is an integer and $1 \leq s \leq w$. Because of the definition of M, we have $t > w$. Setting $x = z$ and $y = t - w$ we get from the triangle inequality

$$z + f(t - w) > f((t - w) + f(z) - 1) = f(t - w + w) = f(t).$$

Hence,

$$f(t-w) \geq f(t) - (z-1) > \frac{z-1}{w}(t-w) + M,$$

a contradiction to the minimality of t.

Therefore the inequality (1) fails for all $t \geq 1$, we have proven

$$f(t) \leq \frac{z-1}{w} \cdot t + M, \qquad (2)$$

instead.

Now, using (2), we finish the proof of Step 3. Because of $z \leq w$ we have $\frac{z-1}{w} < 1$ and we can choose an integer t sufficiently large to fulfill the condition

$$\left(\frac{z-1}{w}\right)^2 t + \left(\frac{z-1}{w}+1\right)M < t.$$

Applying (2) twice we get

$$f(f(t)) \leq \frac{z-1}{w}f(t) + M \leq \frac{z-1}{w}\left(\frac{z-1}{w}t + M\right) + M < t$$

in contradiction to Step 2, which proves Step 3.

Final step. Thus, following Step 2 and Step 3, we obtain

$$z = f(f(z)) \leq f(z) \leq z$$

and $f(z) = z$ for all positive integers z is proven.

6.2.6 Problem 6 – Combinatorics – Russia

First Solution. Let us assume that the statement of the problem was false, and that the sequence $a_1, \ldots, a_n$ of distinct positive integers together with the set M not containing $s = a_1 + \cdots + a_n$ formed a counterexample with n minimum. It is clear that $n \geqslant 2$ and that we may restrict ourselves to the case $a_1 > a_2 > \cdots > a_n$. Setting

$$T_k = a_1 + \cdots + a_k \quad \text{for} \quad k = 0, 1, \ldots, n,$$

we have $0 = T_0 < T_1 < \cdots < T_n = s$. The minimality of n yields

Claim 1. There is no $m \in \{1, 2, \ldots, n\}$ for which the grasshopper is able to perform m jumps without ever landing on a point from M and such that in addition it has succeeded in jumping over at least m elements from M.

Proof. If this was possible with the grasshopper reaching a certain integer t, say, then as $|M \cap (t, s)| \leqslant n - m - 1$ we had $m < n$ and shifting the whole situation by $-t$ we infer from the minimality of n that the grasshopper could also carry the remaining $n - m$ jumps out, contrary to the situation we consider being a counterexample.

Now we call an integer $k \in \{1, 2, \ldots, n\}$ *smooth* if the grasshopper can make k jumps of lengths $a_1, \ldots, a_k$ without ever landing on a point from M, except for the very last jump where it is immaterial whether it lands in M or not. Plainly, 1 is a smooth number and hence there exists a largest integer $k^* \geqslant 1$ for which all of the numbers $1, \ldots, k^*$ are smooth. Obviously $k^* < n$.

Claim 2. We have $T_{k^*} \in M$ and $|M \cap (0, T_{k^*})| \geqslant k^*$.

Proof. Taking a sequence of jumps exemplifying the smoothness of k^* and trying to extend it by appending a jump of length a_{k^*+1} we conclude from the failure of k^*+1 to be smooth that $T_{k^*} \notin M$ is impossible, which proves the first part of our claim. Now assume that $|M \cap (0, T_{k^*})| < k^*$. Then there exists some $\ell \in \{1, \dots, k^*\}$ with $T_{k^*+1} - a_\ell \notin M$ and by the minimality of n the grasshopper can reach $T_{k^*+1} - a_\ell$ starting from the origin by means of jumps of length from $\{a_1, \dots, a_{k^*+1}\} - \{a_\ell\}$, which reveals that k^*+1 is still smooth, which finishes the proof of Claim 2.

What we have shown by now tells us that there is a least $\bar{k} \in \{1, \dots, k^*\}$ with the subtle additional properties $T_{\bar{k}} \in M$ and $|M \cap (0, T_{\bar{k}})| \geqslant \bar{k}$.

Claim 3. The set $M \cap (0, T_{\bar{k}-1}]$ contains fewer than $\bar{k}$ elements.

Proof. This is clear if $\bar{k} = 1$, so suppose $\bar{k} > 1$ from now on. Now, if $T_{\bar{k}-1} \in M$, then our claim follows from the minimal choice of $\bar{k}$, and if $T_{\bar{k}-1} \notin M$, then it follows from the smoothness of $\bar{k}-1$ in conjunction with Claim 1.

Define an integer $v \geqslant 0$ by $|M \cap (0, T_{\bar{k}})| = \bar{k} + v$ and let $r_1 > \dots > r_\ell$ be exactly those indices r from $\{\bar{k}+1, \dots, n\}$ for which $T_{\bar{k}} + a_r \notin M$. Then

$$n - 1 = |M| \geqslant |M \cap (0, T_{\bar{k}})| + 1 + |M \cap (T_{\bar{k}}, s)| \geqslant (\bar{k} + v) + 1 + (n - \bar{k} - \ell),$$

and consequently $\ell \geqslant v + 2$. Note that

$$T_{\bar{k}} + a_{r_1} - a_1, T_{\bar{k}} + a_{r_1} - a_2, \dots, T_{\bar{k}} + a_{r_1} - a_{\bar{k}}, T_{\bar{k}} + a_{r_2} - a_{\bar{k}}, \dots, T_{\bar{k}} + a_{r_{v+2}} - a_{\bar{k}}$$

is an increasing sequence consisting of $\bar{k} + v + 1$ integers below $T_{\bar{k}}$. Using it, we find two indices $r \in \{\bar{k}+1, \dots, n\}$ and $t \in \{1, \dots, \bar{k}\}$ for which $T_{\bar{k}} + a_r \notin M$ and $T_{\bar{k}} + a_r - a_t \notin M$ hold. Now consider the set of jump lengths $B = \{a_1, \dots, a_{\bar{k}}, a_r\} - \{a_t\}$. The sum of these lengths is given by $s' = T_{\bar{k}} + a_r - a_t$ and for this reason it satisfies $s' - \min(B) = T_{\bar{k}} - a_t \leqslant T_{\bar{k}-1}$. By minimality of n and Claim 3, the grasshopper may reach s' by means of the $\bar{k}$ jumps that are contained in B without ever landing on a point from M that is at most $s' - \min(B)$ and thus without ever landing in M at all. From s' it may jump further to $s' + a_t$ and thereby manages in view of the subtle properties of $\bar{k}$ to produce a contradiction to Claim 1, which shows that the assumption we have made at the beginning must have been false, which in turn completes the solution.

Second Solution. We will represent a route of the grasshopper by a sequence of indices $(i_1, \dots, i_n)$ if it makes consecutive jumps $a_{i_1}, \dots, a_{i_n}$. We proceed by induction on n. The cases $n = 1, 2$ are trivial. Now suppose that the claim holds for all values smaller than n. We may assume that $a_1 < \dots < a_n$. Let $d = \min M$.

Case 1. $d < a_n$.

Case 1.1. $a_n \notin M$. Applying the induction hypothesis with a shift by $-a_n$, we obtain that the grasshopper can jump from a_n to s not landing on the points of $M \setminus \{d\}$ using the jumps $a_1, \dots, a_{n-1}$. Putting a_n to the beginning of this sequence of jumps provides a desired route.

Case 1.2. $a_n \in M$. Consider the n pairwise disjoint sets $\{a_n\}, \{a_1, a_1 + a_n\}, \{a_2, a_2 + a_n\}, \dots, \{a_{n-1}, a_{n-1} + a_n\}$. One of them does not intersect M, say $\{a_i, a_i + a_n\}$. Then $|M \cap [a_i + a_n, s]| \leq n - 3$ since $d, a_n < a_i + a_n$. Applying the induction hypothesis with a shift by $-(a_i + a_n)$, we derive that the grasshopper can jump from $a_i + a_n$ to s not landing on the points of M using all jumps except a_i and a_n. Putting a_i, a_n to the beginning of this sequence of jumps provides a desired route.

Case 2. $d \geq a_n$. Let $M' = M \setminus \{d\}$. By the induction hypothesis with a shift by $-a_n$, the grasshopper can jump from a_n to s not landing on the points of M' using the jumps $a_1, \dots, a_{n-1}$. Let the corresponding sequence of indices be $(i_1, \dots, i_{n-1})$.

Case 2.1. The grasshopper does not step on d on this route (including the starting point). Then $d > a_n$. The sequence $U = (n, i_1, \dots, i_{n-1})$ provides a desired route.

Case 2.2. The grasshopper steps on d on this route. Then d is the only point from M that the grasshopper meets on its route. There is some integer k such that $0 \leq k < n-1$ and $a_n + \sum_{j=1}^{k} a_{i_j} = d$. Consider the route represented by $(i_1, \ldots, i_{k+1}, n, i_{k+2}, \ldots, i_{n-1})$. Since $\sum_{j=1}^{k+1} a_{i_j} < \sum_{j=1}^{k} a_{i_j} + a_n = d$, the grasshopper will not land on the points of M during the first $k+1$ jumps. During the remaining part of this route it lands on the same points as in the route U and these points are not less than $\sum_{j=1}^{k+1} a_{i_j} + a_n > d$. Hence it will not land on points of M on this part as well, i.e. on the whole route.

6.3 Awards

6.3.1 Gold Medals - Top Scores

42   Makoto Soejima (Japan)
42   Dongyi Wei (People's Republic of China)
41   Lisa Sauermann (Germany)

6.3.2 Gold Medals

39   Hà Khương Duy (Vietnam)
39   Akio Kishikawa (Japan)
39   Victor Omelyanenko (Russian Federation)
39   Un Song Ri (Democratic People's Republic of Korea)
38   Kazuhiro Hosaka (Japan)
38   Yanlin Zhao (People's Republic of China)
37   Andrew Elvey Price (Australia)
37   Henrique Pondé de Oliveira Pinto (Brazil)
36   Jiaoyang Huang (People's Republic of China)
36   Motoki Takigiku (Japan)
35   John Berman (United States of America)
35   Tak Wing Ching (Hong Kong)
35   Sang Hoon Lee (Republic of Korea)
35   Bo Lin (People's Republic of China)
35   Marsel Matdinov (Russian Federation)
35   Suthee Ruangwises (Thailand)
35   Nazar Serdyuk (Ukraine)
35   István Tomon (Hungary)
35   Konstantin Tyshchuk (Russian Federation)
35   Umut Varolgüneş (Turkey)
35   Bogdan Veklych (Ukraine)
35   Fan Zheng (People's Republic of China)
35   Zhiwei Zheng (People's Republic of China)
34   Hyun Sub Hwang (Republic of Korea)
34   Shiro Imamura (Japan)
34   Jong Chol Kim (Democratic People's Republic of Korea)
34   Süreyya Emre Kurt (Turkey)
34   Eric Larson (United States of America)
34   Elena Mădălina Persu (Romania)
34   Xiaolin (Danny) Shi (Canada)
34   Artsiom Toyestseu (Belarus)
34   Sampson Wong (Australia)
34   Teodor von Burg (Serbia)
33   Omer Cerrahoglu (Romania)
33   Andrea Fogari (Italy)
33   Tae Gu Kang (Republic of Korea)
33   Khashayar Khosravi (Islamic Republic of Iran)
33   Vitaliy Senin (Ukraine)
33   Hsin-Po Wang (Taiwan)

33   Phạm Đức Hùng (Vietnam)
32   Vladimir Bragin (Russian Federation)
32   Tim Hennock (United Kingdom)
32   Gleb Nenashev (Russian Federation)
32   Lyuboslav Panchev (Bulgaria)
32   Giovanni Paolini (Italy)
32   Yong Hyon Ri (Democratic People's Republic of Korea)

6.3.3 Silver Medals

31   Luke Betts (United Kingdom)
31   Pongpak Bhumiwat (Thailand)
31   Germán Stefanich (Argentina)
31   Pietro Vertechi (Italy)
30   Bodan Arsovski (The former Yugoslav Republic of Macedonia)
30   Aaron Wan Yau Chong (Australia)
30   Beka Ergemlidze (Georgia)
30   Amirmasoud Geevechi (Islamic Republic of Iran)
30   János Nagy (Hungary)
30   Evan O'Dorney (United States of America)
30   Nipun Pitimanaaree (Thailand)
30   Kirill Savenkov (Russian Federation)
29   Tae Joo Ahn (Republic of Korea)
29   Sviatlana Auchynnikava (Belarus)
29   Fabio Bioletto (Italy)
29   Mihajlo Cekić (Serbia)
29   Robin Cheng (Canada)
29   Renan Henrique Finder (Brazil)
29   Phạm Hy Hiếu (Vietnam)
29   Fehmi Emre Kadan (Turkey)
29   Supanat Kamtue (Thailand)
29   Palina Khudziakova (Belarus)
29   Tomasz Kociumaka (Poland)
29   Thanard Kurutach (Thailand)
29   Sunkyu Lim (Republic of Korea)
29   Delong Meng (United States of America)
29   Jakub Oćwieja (Poland)
29   Qinxuan Pan (United States of America)
29   Pedro Manuel Passos de Sousa Vieira (Portugal)
29   Jens Reinhold (Germany)
29   Hae Chol Son (Democratic People's Republic of Korea)
29   Chen Sun (Canada)
29   Anibal Velozo (Chile)
28   Tomás Miguel Angles Larico (Peru)
28   Bertram Arnold (Germany)
28   Han Lin Hsieh (Taiwan)
28   Lasha Lakirbaia (Georgia)
28   Cho Ho Lam (Hong Kong)
28   Young Wook Lyoo (Republic of Korea)
28   Julián Alonso Mejía Cordero (Peru)
28   Preeyan Parmar (United Kingdom)

28   Nicolas Radu (Belgium)
28   Kanat Satylkhanov (Kazakhstan)
28   Svetozar Stankov (Bulgaria)
28   Pooya Vahidi Ferdowsi (Islamic Republic of Iran)
28   Aliaksei Vaidzelevich (Belarus)
28   Chien-Yi Wang (Taiwan)
28   Zhivko Zhechev (Bulgaria)
28   Loke Zhi Kin (Malaysia)
28   András Éles (Hungary)
28   Melih Üçer (Turkey)
27   Jang Su Choe (Democratic People's Republic of Korea)
27   Andrei Deneanu (Romania)
27   Pakawut Jiradilok (Thailand)
27   Vaidotas Juronis (Lithuania)
27   Nina Kamčev (Croatia)
27   Nikoloz Machavariani (Georgia)
27   Dušan Milijančević (Serbia)
27   Luka Milićević (Serbia)
27   Yu-Fan Tung (Taiwan)
26   Vefa Göksel (Turkey)
26   Svetoslav Karaivanov (Bulgaria)
26   Yegor Klochkov (Kazakhstan)
26   Ricardo Jesús Ramos Castillo (Peru)
26   Altun Shukurlu (Azerbaijan)
26   Hunter Spink (Canada)
26   Marcelo Tadeu de Sá Oliveira Sales (Brazil)
25   Christos Anastassiades (Cyprus)
25   Wenyu Cao (United States of America)
25   Jia-Han Chiam (Singapore)
25   César Cuenca Lucero (Peru)
25   Nguyễn Hoàng Hải (Vietnam)
25   Ufuk Kanat (Turkey)
25   Mathias Tejs Knudsen (Denmark)
25   Malte Lackmann (Germany)
25   Peter Leach (United Kingdom)
25   Martin Merker (Germany)
25   Josef Tkadlec (Czech Republic)
25   Sameer Wagh (India)
25   Chen-Yu Yang (Taiwan)
24   Wouter Berkelmans (Netherlands)
24   Francisc Bozgan (Romania)
24   Hossein Dabirian (Islamic Republic of Iran)
24   Nicolás Del Castillo (Colombia)
24   Sergey Dovgal (Belarus)
24   Akashnil Dutta (India)
24   Sepehr Ghazi Nezami (Islamic Republic of Iran)
24   Ka Kin Kenneth Hung (Hong Kong)
24   Nursultan Khadjimuratov (Kazakhstan)
24   Yao Chen Ivan Loh (Singapore)
24   Anastasiya Lysakevych (Ukraine)
24   Jean-François Martin (France)
24   Inomzhon Mirzaev (Tajikistan)
24   Diyora Salimova (Uzbekistan)
24   Agageldi Samedov (Turkmenistan)

24   Matheus Secco Torres da Silva (Brazil)
24   Ananth Shankar (India)
24   Yi-Chan Wu (Taiwan)

6.3.4 Bronze Medals

23   Liam Baker (South Africa)
23   Thomas Budzinski (France)
23   Subhadip Chowdhury (India)
23   Luca Ghidelli (Italy)
23   Andreas Dwi Maryanto Gunawan (Indonesia)
23   Artsiom Hovarau (Belarus)
23   Suguru Ishikawa (Japan)
23   Kristóf Kornis (Hungary)
23   Christoph Kröner (Germany)
23   Jorge Ricardo Landeira da Silva Miranda (Portugal)
23   Ambroise Marigot (France)
23   Dániel Nagy (Hungary)
23   Dimitrios Papadimitriou (Greece)
23   Gaurav Digambar Patil (India)
23   Tudor Pădurariu (Romania)
23   Tolebi Sailauov (Kazakhstan)
23   Galin Statev (Bulgaria)
23   Vukašin Stojisavljević (Serbia)
23   Chengyue (Jarno) Sun (Canada)
23   Levan Varamashvili (Georgia)
22   Seyed Ehsan Azarmsa (Islamic Republic of Iran)
22   Pēteris Eriņš (Latvia)
22   Ho Gon Jon (Democratic People's Republic of Korea)
22   Karol Konaszyński (Poland)
22   Stacey Wing Chee Law (Australia)
22   Lajanugan Logeswaran (Sri Lanka)
22   Davi Lopes Alves De Medeiros (Brazil)
22   Marco Antonio Lopes Pedroso (Brazil)
22   Damian Orlef (Poland)
22   Tomasz Pawłowski (Poland)
22   Lasha Peradze (Georgia)
22   Ilgar Ramazanli (Azerbaijan)
22   Marius Tiba (Romania)
21   Ryan Jun Neng Chan (Singapore)
21   Enkhzaya Enkhtaivan (Mongolia)
21   Reynaldo Gil Pons (Cuba)
21   Percy Augusto Guerra Ríos (Peru)
21   Igor Kudla (Ukraine)
21   Sean Moss (United Kingdom)
21   Ohad Nir (Israel)
21   Kakamyrat Tushiyev (Turkmenistan)
21   Raymond van Bommel (Netherlands)
20   Chris Bellin (United Kingdom)
20   Michal Hagara (Slovakia)
20   Moisés Herradón Cueto (Spain)

20   Gaygysyz Hojanazarov (Turkmenistan)
20   Greta Kuprijanovaitė (Lithuania)
20   Aldrian Obaja Muis (Indonesia)
20   Victor Valov (Bulgaria)
20   Goran Žužić (Croatia)
19   Ivo Božić (Croatia)
19   Nazar Emirov (Turkmenistan)
19   Eben Freeman (Switzerland)
19   Andrei Ivanov (Republic of Moldova)
19   Sondre Kvamme (Norway)
19   Florian Meier (Liechtenstein)
19   Muhamadjon Shoev (Tajikistan)
19   Mats Vermeeren (Belgium)
19   Sergio Véga (France)
19   Tạ Đức Thành (Vietnam)
18   Vahagn Aslanyan (Armenia)
18   Jorge Francisco Barreras Cortes (Colombia)
18   Omri Ben Eliezer (Israel)
18   Andrei Iliaşenco (Republic of Moldova)
18   Adrian Satja Kurdija (Croatia)
18   Jeck Lim (Singapore)
18   Jingran Lin (Luxembourg)
18   Jorge Alberto Olarte (Colombia)
18   Sanzhar Orazbayev (Kazakhstan)
18   Rafael Ángel Rodríguez Arguedas (Costa Rica)
18   Raphael Steiner (Switzerland)
18   Gergely Szűcs (Hungary)
17   Ping Ngai Chung (Hong Kong)
17   Felix Dräxler (Austria)
17   Vahagn Kirakosyan (Armenia)
17   Pappelis Konstantinos (Greece)
17   Kirill Kuzmin (Italy)
17   Alfred Liang (Australia)
17   Birzhan Muldagaliyev (Kazakhstan)
17   Zorigoo Ochirkhuyag (Mongolia)
17   Samin Riasat (Bangladesh)
17   Iván Sadofschi (Argentina)
17   Jonathan Schneider (Canada)
17   Chi Choi Wong (Macau)
17   Peter Zarén (Sweden)
16   Martin Bachratý (Slovakia)
16   César Bibiano Velasco (Mexico)
16   Raúl Arturo Chávez Sarmiento (Peru)
16   Erik Alejandro Gallegos Baños (Mexico)
16   Buddhima Ruwanmini Gamlath Gamlath Ralalage (Sri Lanka)
16   Malcolm Granville (New Zealand)
16   Emil Jafarli (Azerbaijan)
16   Bojan Joveski (The former Yugoslav Republic of Macedonia)
16   Alexander Lemmens (Belgium)
16   Fotis Logothetis (Greece)
16   Manuel Guillermo López Buenfil (Mexico)
16   Azizkhon Nazarov (Uzbekistan)
16   Niels Olsen (Denmark)
16   Abror Pirnapasov (Uzbekistan)

16	Jakub Witaszek (Poland)
16	Nguyễn Xuân Cương (Vietnam)
15	Joseph Andreas (Indonesia)
15	Glenier Lázaro Bello Burguet (Spain)
15	Ricardo Correia Moreira (Portugal)
15	Iván Geffner (Spain)
15	Ofir Gorodetsky (Israel)
15	Dragoş Gîlcă (Republic of Moldova)
15	Pierre Haas (Luxembourg)
15	Ander Lamaison (Spain)
15	Carmela Antoinette Sio Lao (Philippines)
15	Matko Ljulj (Croatia)
15	Tian Wen Daniel Low (Singapore)
15	Jan Matějka (Czech Republic)
15	João Morais Carreira Pereira (Portugal)
15	Stefan Stojchevski (The former Yugoslav Republic of Macedonia)
15	Henry Thackeray (South Africa)
14	Matej Aleksandrov (Slovenia)
14	Luvsanbyamba Buyankhuu (Mongolia)
14	Nazia Naser Chowdhury (Bangladesh)
14	Hrvoje Dujmovic (Switzerland)
14	Komron Giesiev (Tajikistan)
14	Alexandru Grigoroi (Republic of Moldova)
14	Predrag Gruevski (The former Yugoslav Republic of Macedonia)
14	Johannes Hafner (Austria)
14	Gabriel Isheden (Sweden)
14	Beyrem Khalfaoui (Tunisia)
14	José Ramón Madrid Padilla (Honduras)
14	Ngai Fung Ng (Hong Kong)
14	Jelena Radović (Bosnia and Herzegovina)
14	Darya Shchedrina (Ukraine)
14	Marc Sinner (Luxembourg)
14	Jarle Stavnes (Norway)
14	Waqar Ali Syed (Pakistan)
14	Jan Vaňhara (Czech Republic)
14	Ronald Widjojo (Indonesia)

6.3.5 Honourable Mentions

13	Adnan Ademović (Bosnia and Herzegovina)
13	Flavio Hernández González (Mexico)
13	Kairi Kangro (Estonia)
13	Linas Klimavičius (Lithuania)
13	Nikola Milinkovic (Montenegro)
13	Tarik Adnan Moon (Bangladesh)
13	Stephan Pfannerer (Austria)
13	Subhan Rustamli (Azerbaijan)
13	Rauno Siinmaa (Estonia)
13	Jia Hao Barry Tng (Singapore)
12	Ibrokhimbek Akramov (Uzbekistan)
12	Elvin Aliyev (Azerbaijan)

12	George Arzeno (Puerto Rico)
12	Hampus Engsner (Sweden)
12	Antonio Fan (Panama)
12	Mukundadura Yasod Sankalpa Fonseka (Sri Lanka)
12	Jean Garcin (France)
12	Arlton Gilbert (South Africa)
12	David Klaška (Czech Republic)
12	Aleksis Koski (Finland)
12	Matjaž Leonardis (Slovenia)
12	Miguel Maurizio (Argentina)
12	Rickard Norlander (Sweden)
12	Ieva Ozola (Latvia)
12	Heikki Pulkkinen (Finland)
12	Sean Wentzel (South Africa)
11	Angela Castañeda (Colombia)
11	Anthony Santiago Chaves Aguilar (Costa Rica)
11	Borna Cicvarić (Croatia)
11	Francois Conradie (South Africa)
11	Gregor Grasselli (Slovenia)
11	Greg Jackson (South Africa)
11	Inbar Klang (Israel)
11	David Kok (Netherlands)
11	Anja Komatar (Slovenia)
11	Dana Ma (Australia)
11	Sébastien Miquel (France)
11	Akshay Mittal (India)
11	Chandresh Amrit Ramlagan (Trinidad and Tobago)
11	Filip Sládek (Slovakia)
11	Evangelos Taratoris (Greece)
11	Jakub Uhrík (Slovakia)
11	Samuel Říha (Czech Republic)
10	Admir Beširević (Bosnia and Herzegovina)
10	Carlos Carvajal (Colombia)
10	Ratko Darda (Bosnia and Herzegovina)
10	Gregory Debruyne (Belgium)
10	Ilias Giechaskiel (Greece)
10	Haque Muhammad Ishfaq (Bangladesh)
10	Chi Tou Lam (Macau)
10	Josef Ondřej (Czech Republic)
10	Clemens Pohle (Switzerland)
10	David Shin (New Zealand)
10	Kerry Shastri Singh (Trinidad and Tobago)
10	Tsotne Tabidze (Georgia)
10	Filip Talimdzioski (The former Yugoslav Republic of Macedonia)
10	Andre Tamm (Estonia)
9	Sivert Bocianowski (Norway)
9	Harm Campmans (Netherlands)
9	Gregoire Genest (Luxembourg)
9	Yuval Goldberg (Israel)
9	Karl Erik Holter (Norway)
9	Nikolaj Kammersgaard (Denmark)
9	Pranon Rahman Khan (Bangladesh)
9	Kan Chun Leong (Macau)
9	Clemens Müllner (Austria)

9   Gonçalo Pereira Simões Matos (Portugal)
9   Xavier Soriano (Ecuador)
9   Alejandro José Vargas De León (Guatemala)
9   Don Anton Tharindu Kumar Warnakula Warnakulaarachchiralalage (Sri Lanka)
9   Ilias Zadik (Greece)
9   Ariel Zylber (Argentina)
8   Michael Anastos (Cyprus)
8   Peter Csiba (Slovakia)
8   Matias Escuder (Uruguay)
8   Jenny Johansson (Sweden)
8   Paulius Kantautas (Lithuania)
8   Ben Kornfeld (New Zealand)
8   Raja Oktovin Parhasian Damanik (Indonesia)
8   Raúl Queiroz Do Vale de Noronha Penaguião (Portugal)
8   Hayk Saribekyan (Armenia)
8   Pascal Su (Switzerland)
8   Lasse Vekama (Finland)
8   Normunds Vilciņš (Latvia)
8   Azamat Zhabykeev (Kyrgyzstan)
7   Susarbek Baibaev (Kyrgyzstan)
7   Ingólfur Eðvarðsson (Iceland)
7   Mahamarakkalage Dileepa Yasas Fernando (Sri Lanka)
7   Ilkhomzhon Kalandarov (Kyrgyzstan)
7   Eric Larsson (Sweden)
7   Byambadorj Otgonsuren (Mongolia)
7   Georgios Panagopoulos (Cyprus)
7   Edvard Poliakov (Lithuania)
7   Ha Young Shin (New Zealand)
7   Jānis Smilga (Latvia)
7   Topi Talvitie (Finland)

6.4 Individual Scores

Albania   ALB

ALB-1		9	1 3 0 4 1 0	Andi Reçi
ALB-2		2	1 1 0 0 0 0	Tedi Aliaj
ALB-3		3	1 2 0 0 0 0	Ornela Xhelili
ALB-4		0	0 0 0 0 0 0	Arlind Gjoka
ALB-5		5	2 1 0 1 1 0	Niko Kaso
ALB-6		5	4 0 0 0 1 0	Ridgers Mema

Algeria   ALG

ALG-1		1	0 1 0 0 0 0	Lamia Attouche
ALG-2		1	1 0 0 0 0 0	Hacen Zelaci
ALG-3		0	0 0 0 0 0 0	Oussama Guessab
ALG-4		0	0 0 0 0 0 0	Kacem Hariz

Argentina   ARG

ARG-1	**S**	31	7 7 7 3 7 0	Germán Stefanich
ARG-2		11	2 3 0 0 6 0	Alfredo Umfurer
ARG-3	**B**	17	7 7 1 2 0 0	Iván Sadofschi
ARG-4		13	6 1 1 0 5 0	Federico Cogorno
ARG-5	**HM**	9	7 1 0 1 0 0	Ariel Zylber
ARG-6	**HM**	12	7 2 0 1 2 0	Miguel Maurizio

Armenia   ARM

ARM-1		9	4 5 0 0 0 0	Anna Srapionyan
ARM-2		1	1 0 0 0 0 0	Nerses Srapionyan
ARM-3	**B**	17	7 5 0 5 0 0	Vahagn Kirakosyan
ARM-4	**B**	18	7 6 0 4 1 0	Vahagn Aslanyan
ARM-5	**HM**	8	7 0 0 1 0 0	Hayk Saribekyan
ARM-6		6	0 0 0 5 1 0	Vanik Tadevosyan

Australia   AUS

AUS-1	**S**	30	7 7 7 2 7 0	Aaron Wan Yau Chong
AUS-2	**G**	37	7 6 7 5 7 5	Andrew Elvey Price
AUS-3	**B**	22	5 7 0 7 3 0	Stacey Wing Chee Law
AUS-4	**B**	17	7 3 0 0 7 0	Alfred Liang
AUS-5	**HM**	11	7 4 0 0 0 0	Dana Ma
AUS-6	**G**	34	7 7 7 7 6 0	Sampson Wong

Austria   AUT

AUT-1	**B**	17	7 2 0 5 3 0	Felix Dräxler
AUT-2		7	4 1 0 0 2 0	Adrian Fuchs
AUT-3	**B**	14	5 2 0 0 7 0	Johannes Hafner
AUT-4	**HM**	9	7 1 0 0 1 0	Clemens Müllner
AUT-5	**HM**	13	1 7 0 5 0 0	Stephan Pfannerer
AUT-6		6	2 1 0 3 0 0	Valerie Roitner

Azerbaijan   AZE

AZE-1	**B**	22	7 2 0 6 7 0	Ilgar Ramazanli
AZE-2	**HM**	12	7 0 0 5 0 0	Elvin Aliyev
AZE-3		2	1 1 0 0 0 0	Zulfu Aslanli
AZE-4	**S**	26	7 7 0 5 7 0	Altun Shukurlu
AZE-5	**HM**	13	0 7 0 6 0 0	Subhan Rustamli
AZE-6	**B**	16	2 7 0 7 0 0	Emil Jafarli

Bangladesh   BGD

BGD-1	**B**	17	7 7 0 1 2 0	Samin Riasat
BGD-2	**B**	14	1 7 0 6 0 0	Nazia Naser Chowdhury
BGD-3	**HM**	13	1 7 0 5 0 0	Tarik Adnan Moon
BGD-4	**HM**	10	2 7 0 1 0 0	Haque Muhammad Ishfaq
BGD-5		4	3 0 0 1 0 0	Kazi Hasan Zubaer
BGD-6	**HM**	9	7 1 0 1 0 0	Pranon Rahman Khan

Belarus   BLR

BLR-1	**S**	29	2 7 7 6 7 0	Sviatlana Auchynnikava
BLR-2	**S**	24	7 4 0 6 7 0	Sergey Dovgal
BLR-3	**B**	23	7 2 0 7 7 0	Artsiom Hovarau
BLR-4	**S**	29	7 7 1 7 7 0	Palina Khudziakova
BLR-5	**G**	34	7 7 7 6 7 0	Artsiom Toyestseu
BLR-6	**S**	28	7 7 0 7 7 0	Aliaksei Vaidzelevich

Belgium   BEL

BEL-1		13	6 0 0 5 2 0	Loïc Burger
BEL-2		3	2 0 0 1 0 0	Cédric De Groote
BEL-3	**S**	28	7 7 2 5 7 0	Nicolas Radu
BEL-4	**HM**	10	7 2 0 1 0 0	Gregory Debruyne
BEL-5	**B**	16	7 1 1 7 0 0	Alexander Lemmens
BEL-6	**B**	19	7 7 0 5 0 0	Mats Vermeeren

Benin   BEN

| BEN-1 | | 2 | 0 1 0 0 1 0 | Arélyss Eblohoué |
| BEN-2 | | 1 | 1 0 0 0 0 0 | Comlan Edmond Koudjinan |

Bolivia   BOL

BOL-1		1	0 0 0 1 0 0	Erick Daniel Vicente Minaya
BOL-3		7	3 3 0 1 0 0	Diego Salazar Gutiérrez
BOL-4		1	0 1 0 0 0 0	Álvaro Rubén Hurtado Maldonado

Bosnia and Herzegovina   BIH

BIH-1	HM	13	7 3 0 0 3 0	Adnan Ademović
BIH-2	HM	10	7 3 0 0 0 0	Admir Beširević
BIH-3	B	14	5 1 0 5 3 0	Jelena Radović
BIH-4		10	4 0 0 6 0 0	Vlado Uljarević
BIH-5	HM	10	7 1 0 2 0 0	Ratko Darda
BIH-6		6	3 0 0 3 0 0	Mina Ferizbegović

Brazil   BRA

BRA-1	G	37	7 7 7 6 7 3	Henrique Pondé de Oliveira Pinto
BRA-2	S	29	7 7 1 7 7 0	Renan Henrique Finder
BRA-3	S	26	7 5 0 7 7 0	Marcelo Tadeu de Sá Oliveira Sales
BRA-4	B	22	7 6 1 5 3 0	Davi Lopes Alves De Medeiros
BRA-5	B	22	7 5 1 6 3 0	Marco Antonio Lopes Pedroso
BRA-6	S	24	7 7 0 7 3 0	Matheus Secco Torres da Silva

Bulgaria   BGR

BGR-1	G	32	7 7 7 4 7 0	Lyuboslav Panchev
BGR-2	S	28	7 7 2 5 7 0	Svetozar Stankov
BGR-3	S	28	7 7 0 7 7 0	Zhivko Zhechev
BGR-4	B	23	7 3 0 6 7 0	Galin Statev
BGR-5	S	26	7 7 0 5 7 0	Svetoslav Karaivanov
BGR-6	B	20	7 7 0 5 1 0	Victor Valov

Cambodia   KHM

KHM-1		6	2 3 0 1 0 0	Sopheak Touch
KHM-2		1	0 1 0 0 0 0	Kunthea Din
KHM-3		1	0 1 0 0 0 0	Leanghour Lim
KHM-4		1	0 0 0 1 0 0	Guechlaing Chea
KHM-5		2	0 1 0 1 0 0	Panha Ouk
KHM-6		3	2 1 0 0 0 0	Nam Seang

Canada   CAN

CAN-1	S	29	7 6 3 6 7 0	Robin Cheng
CAN-2	B	17	6 4 1 5 1 0	Jonathan Schneider
CAN-3	G	34	7 7 6 7 7 0	Xiaolin (Danny) Shi
CAN-4	S	26	7 4 1 7 7 0	Hunter Spink
CAN-5	S	29	7 7 1 7 7 0	Chen Sun
CAN-6	B	23	7 7 1 1 7 0	Chengyue (Jarno) Sun

Chile   CHI

CHI-1	S	29	7 7 7 5 3 0	Anibal Velozo
CHI-2		1	0 1 0 0 0 0	Benjamin Baeza
CHI-3		11	3 2 0 6 0 0	Sebastian Zuñiga
CHI-4		0	0 0 0 0 0 0	Mauricio Garcia

People's Republic of China   CHN

CHN-1	G	42	7 7 7 7 7 7	Dongyi Wei
CHN-2	G	35	7 7 7 7 7 0	Bo Lin
CHN-3	G	35	7 7 7 7 7 0	Fan Zheng
CHN-4	G	35	7 7 7 7 7 0	Zhiwei Zheng
CHN-5	G	38	7 7 7 7 7 3	Yanlin Zhao
CHN-6	G	36	7 7 7 7 7 1	Jiaoyang Huang

Colombia   COL

COL-1	B	18	7 3 7 0 1 0	Jorge Alberto Olarte
COL-2	HM	11	7 1 1 0 2 0	Angela Castañeda
COL-3		7	1 2 1 1 2 0	Hayden Liu Weng
COL-4	B	18	7 7 0 4 0 0	Jorge Francisco Barreras Cortes
COL-5	S	24	7 7 1 6 3 0	Nicolás Del Castillo
COL-6	HM	10	7 1 0 0 2 0	Carlos Carvajal

Costa Rica   CRI

CRI-1		5	2 1 0 2 0 0	Christopher Antonio Trejos Castillo
CRI-2	B	18	7 2 1 6 2 0	Rafael Ángel Rodríguez Arguedas
CRI-3	HM	11	7 3 0 1 0 0	Anthony Santiago Chaves Aguilar
CRI-4		0	0 0 0 0 0 0	Ezequiel Heredia Fernández

Croatia   HRV

HRV-1	**B**	19	7 7 0 3 2 0	Ivo Božić
HRV-2	**HM**	11	7 2 1 0 1 0	Borna Cicvarić
HRV-3	**S**	27	7 7 1 7 5 0	Nina Kamčev
HRV-4	**B**	18	4 7 0 7 0 0	Adrian Satja Kurdija
HRV-5	**B**	15	7 7 0 1 0 0	Matko Ljulj
HRV-6	**B**	20	7 7 1 2 3 0	Goran Žužić

Cuba   CUB

| CUB-1 | **B** | 21 | 7 6 1 5 2 0 | Reynaldo Gil Pons |

Cyprus   CYP

CYP-1	**HM**	8	7 0 0 0 1 0	Michael Anastos
CYP-2	**S**	25	7 7 1 7 3 0	Christos Anastassiades
CYP-3	**HM**	7	7 0 0 0 0 0	Georgios Panagopoulos
CYP-4		1	1 0 0 0 0 0	Talia Tseriotou
CYP-5		1	0 1 0 0 0 0	Maria Michaelidou
CYP-6		3	1 2 0 0 0 0	Neofytos Apostolou

Czech Republic   CZE

CZE-1	**HM**	12	7 1 0 0 4 0	David Klaška
CZE-2	**B**	15	6 6 1 0 2 0	Jan Matějka
CZE-3	**HM**	10	7 3 0 0 0 0	Josef Ondřej
CZE-4	**HM**	11	7 3 0 1 0 0	Samuel Říha
CZE-5	**S**	25	7 7 1 7 3 0	Josef Tkadlec
CZE-6	**B**	14	6 0 1 0 7 0	Jan Vaňhara

Denmark   DEN

DEN-1	**S**	25	7 7 0 4 7 0	Mathias Tejs Knudsen
DEN-2	**HM**	9	7 2 0 0 0 0	Nikolaj Kammersgaard
DEN-3		1	1 0 0 0 0 0	Ben Braithwaite
DEN-4	**B**	16	7 5 1 1 2 0	Niels Olsen
DEN-5		11	6 0 2 0 0 3	Rasmus Nørtoft Johansen
DEN-6		6	5 0 0 1 0 0	Pernille Hanehøj

Ecuador   ECU

ECU-1	**HM**	9	1 1 0 7 0 0	Xavier Soriano
ECU-2		2	0 2 0 0 0 0	Gabriel Bravo
ECU-3		9	6 0 0 3 0 0	Christian Abad

ECU-4		3	2 1 0 0 0 0	Miguel Ordoñez
ECU-5		0	0 0 0 0 0 0	Christian Pihuave
ECU-6		3	0 3 0 0 0 0	Caril Martinez

Estonia   EST

EST-1	HM	13	7 4 0 0 2 0	Kairi Kangro
EST-2		1	1 0 0 0 0 0	Heino Soo
EST-3		2	1 1 0 0 0 0	Paavo Parmas
EST-4	HM	13	7 2 0 4 0 0	Rauno Siinmaa
EST-5		1	0 1 0 0 0 0	Aleksandr Šved
EST-6	HM	10	7 0 1 2 0 0	Andre Tamm

Finland   FIN

FIN-1	HM	12	7 0 0 2 3 0	Aleksis Koski
FIN-2		7	5 0 0 2 0 0	Konsta Lensu
FIN-3	HM	12	4 1 0 0 7 0	Heikki Pulkkinen
FIN-4		3	1 0 0 2 0 0	Alexey Sofiev
FIN-5	HM	7	7 0 0 0 0 0	Topi Talvitie
FIN-6	HM	8	7 1 0 0 0 0	Lasse Vekama

France   FRA

FRA-1	B	23	7 7 2 7 0 0	Thomas Budzinski
FRA-2	HM	12	7 2 0 3 0 0	Jean Garcin
FRA-3	B	23	5 6 1 4 7 0	Ambroise Marigot
FRA-4	S	24	7 6 0 4 7 0	Jean-François Martin
FRA-5	HM	11	7 1 1 0 2 0	Sébastien Miquel
FRA-6	B	19	7 2 3 0 7 0	Sergio Véga

Georgia   GEO

GEO-1	S	30	7 7 3 6 7 0	Beka Ergemlidze
GEO-2	S	28	7 7 1 6 7 0	Lasha Lakirbaia
GEO-3	S	27	7 7 1 6 6 0	Nikoloz Machavariani
GEO-4	B	22	7 7 0 5 3 0	Lasha Peradze
GEO-5	HM	10	7 1 1 1 0 0	Tsotne Tabidze
GEO-6	B	23	7 2 6 5 3 0	Levan Varamashvili

Germany   GER

GER-1	S	28	7 7 0 7 7 0	Bertram Arnold
GER-2	B	23	7 7 0 7 2 0	Christoph Kröner
GER-3	S	25	7 3 2 6 7 0	Malte Lackmann

GER-4	**S**	25	7 3 1 4 7 3	Martin Merker
GER-5	**S**	29	7 7 1 7 7 0	Jens Reinhold
GER-6	**G**	41	7 6 7 7 7 7	Lisa Sauermann

Greece   HEL

HEL-1	**HM**	10	7 1 0 1 1 0	Ilias Giechaskiel
HEL-2	**B**	16	7 3 0 6 0 0	Fotis Logothetis
HEL-3	**B**	23	7 7 0 5 4 0	Dimitrios Papadimitriou
HEL-4	**B**	17	4 7 0 5 1 0	Pappelis Konstantinos
HEL-5	**HM**	11	4 7 0 0 0 0	Evangelos Taratoris
HEL-6	**HM**	9	7 1 0 0 1 0	Ilias Zadik

Guatemala   GTM

GTM-1		1	1 0 0 0 0 0	Francisco José Martínez Figueroa
GTM-2		3	0 3 0 0 0 0	Marcos Fernando Galindo Gomez
GTM-3	**HM**	9	7 0 0 2 0 0	Alejandro José Vargas De León
GTM-4		1	0 0 0 0 1 0	José Carlos Arandi Ayala

Honduras   HND

HND-1		4	3 1 0 0 0 0	Nestor Alejandro Bermudez
HND-2		6	3 3 0 0 0 0	Sergio David Manzanarez
HND-3	**B**	14	1 7 0 6 0 0	José Ramón Madrid Padilla

Hong Kong   HKG

HKG-1	**G**	35	7 7 7 7 7 0	Tak Wing Ching
HKG-2	**B**	17	3 7 1 6 0 0	Ping Ngai Chung
HKG-3	**S**	28	7 6 1 7 7 0	Cho Ho Lam
HKG-4	**S**	24	7 7 1 7 2 0	Ka Kin Kenneth Hung
HKG-5	**B**	14	1 7 0 5 1 0	Ngai Fung Ng
HKG-6		4	0 1 0 0 3 0	Tak Hei Yu

Hungary   HUN

HUN-1	**S**	28	7 7 2 5 7 0	András Éles
HUN-2	**B**	23	7 7 0 2 7 0	Kristóf Kornis
HUN-3	**B**	23	7 7 1 5 3 0	Dániel Nagy
HUN-4	**S**	30	7 7 7 2 7 0	János Nagy
HUN-5	**B**	18	7 7 1 0 3 0	Gergely Szűcs
HUN-6	**G**	35	7 7 7 7 7 0	István Tomon

Iceland   ISL

ISL-1		5	5 0 0 0 0 0	Arna Pálsdóttir
ISL-2		0	0 0 0 0 0 0	Helga Kristín Ólafsdóttir
ISL-3		6	6 0 0 0 0 0	Ögmundur Eiríksson
ISL-4		4	3 0 0 0 1 0	Helgi Kristjánsson
ISL-5	HM	7	7 0 0 0 0 0	Ingólfur Eðvarðsson
ISL-6		4	4 0 0 0 0 0	Paul Frigge

India   IND

IND-1	S	24	7 0 1 5 7 4	Akashnil Dutta
IND-2	HM	11	7 1 1 2 0 0	Akshay Mittal
IND-3	B	23	7 7 3 0 6 0	Gaurav Digambar Patil
IND-4	S	24	7 5 0 5 7 0	Ananth Shankar
IND-5	S	25	7 7 1 3 7 0	Sameer Wagh
IND-6	B	23	7 6 1 6 3 0	Subhadip Chowdhury

Indonesia   IDN

IDN-1	B	15	6 2 0 6 1 0	Joseph Andreas
IDN-2	B	20	7 0 1 5 7 0	Aldrian Obaja Muis
IDN-3	B	23	7 2 0 7 7 0	Andreas Dwi Maryanto Gunawan
IDN-4	HM	8	7 1 0 0 0 0	Raja Oktovin Parhasian Damanik
IDN-5		4	1 1 0 1 1 0	Ivan Wangsa Cipta Lingga
IDN-6	B	14	7 7 0 0 0 0	Ronald Widjojo

Islamic Republic of Iran   IRN

IRN-1	B	22	7 7 1 0 7 0	Seyed Ehsan Azarmsa
IRN-2	S	24	7 7 1 6 3 0	Hossein Dabirian
IRN-3	S	30	7 7 4 5 7 0	Amirmasoud Geevechi
IRN-4	S	24	7 7 1 6 3 0	Sepehr Ghazi Nezami
IRN-5	G	33	7 7 2 7 7 3	Khashayar Khosravi
IRN-6	S	28	7 7 1 6 7 0	Pooya Vahidi Ferdowsi

Ireland   IRL

IRL-1		6	2 2 0 2 0 0	Jack McKenna
IRL-2		4	3 1 0 0 0 0	David McCarthy
IRL-3		2	0 1 0 1 0 0	Colman Humphrey
IRL-4		4	1 1 0 2 0 0	Colin Egan
IRL-5		2	0 0 0 2 0 0	Vicki McAvinue
IRL-6		2	1 1 0 0 0 0	Cillian Power

Israel   ISR

ISR-1	**B**	18	4 5 2 0 7 0	Omri Ben Eliezer
ISR-2	**B**	15	7 1 1 6 0 0	Ofir Gorodetsky
ISR-3	**B**	21	7 7 0 7 0 0	Ohad Nir
ISR-4	**HM**	11	7 2 1 1 0 0	Inbar Klang
ISR-5		6	4 2 0 0 0 0	Rom Dudkiewicz
ISR-6	**HM**	9	7 1 0 0 1 0	Yuval Goldberg

Italy   ITA

ITA-1	**S**	29	7 7 1 7 7 0	Fabio Bioletto
ITA-2	**G**	33	7 7 5 7 7 0	Andrea Fogari
ITA-3	**B**	23	7 7 1 5 3 0	Luca Ghidelli
ITA-4	**B**	17	7 3 0 0 7 0	Kirill Kuzmin
ITA-5	**G**	32	7 7 4 7 7 0	Giovanni Paolini
ITA-6	**S**	31	7 7 7 3 7 0	Pietro Vertechi

Japan   JPN

JPN-1	**G**	38	7 7 7 7 7 3	Kazuhiro Hosaka
JPN-2	**G**	34	7 7 7 6 7 0	Shiro Imamura
JPN-3	**B**	23	7 7 1 1 7 0	Suguru Ishikawa
JPN-4	**G**	39	7 7 7 7 7 4	Akio Kishikawa
JPN-5	**G**	42	7 7 7 7 7 7	Makoto Soejima
JPN-6	**G**	36	7 7 5 5 7 5	Motoki Takigiku

Kazakhstan   KAZ

KAZ-1	**S**	26	7 6 1 5 7 0	Yegor Klochkov
KAZ-2	**B**	18	7 3 0 5 3 0	Sanzhar Orazbayev
KAZ-3	**S**	24	7 7 1 7 2 0	Nursultan Khadjimuratov
KAZ-4	**S**	28	7 7 1 6 7 0	Kanat Satylkhanov
KAZ-5	**B**	17	7 7 0 3 0 0	Birzhan Muldagaliyev
KAZ-6	**B**	23	7 7 1 5 3 0	Tolebi Sailauov

Democratic People's Republic of Korea   PRK

PRK-1	**G**	39	7 7 7 7 7 4	Un Song Ri
PRK-2	**B**	22	7 2 1 7 2 3	Ho Gon Jon
PRK-3	**G**	34	7 7 7 7 6 0	Jong Chol Kim
PRK-4	**G**	32	7 7 7 4 7 0	Yong Hyon Ri
PRK-5	**S**	27	7 5 1 7 7 0	Jang Su Choe
PRK-6	**S**	29	7 7 1 7 7 0	Hae Chol Son

Republic of Korea   KOR

KOR-1	**G**	33	7 7 5 7 7 0	Tae Gu Kang
KOR-2	**S**	28	7 7 1 6 7 0	Young Wook Lyoo
KOR-3	**S**	29	7 7 1 7 7 0	Tae Joo Ahn
KOR-4	**G**	35	7 7 7 7 7 0	Sang Hoon Lee
KOR-5	**S**	29	7 7 1 7 7 0	Sunkyu Lim
KOR-6	**G**	34	7 7 7 6 7 0	Hyun Sub Hwang

Kuwait   KWT

KWT-1	1	0 1 0 0 0 0	Ali Ahmad Nawab
KWT-2	1	0 1 0 0 0 0	Ahmad Ameen Alshemali
KWT-4	1	0 1 0 0 0 0	Hessah Abdullatif Albanwan
KWT-5	0	0 0 0 0 0 0	Hadeel Awadh Abdullah

Kyrgyzstan   KGZ

KGZ-1	**HM**	8	1 7 0 0 0 0	Azamat Zhabykeev
KGZ-2		1	0 1 0 0 0 0	Radik Srajidinov
KGZ-3	**HM**	7	0 7 0 0 0 0	Ilkhomzhon Kalandarov
KGZ-4		10	1 3 0 6 0 0	Azamat Askarov
KGZ-5	**HM**	7	0 7 0 0 0 0	Susarbek Baibaev
KGZ-6		0	0 0 0 0 0 0	Meerim Topchubaeva

Latvia   LVA

LVA-1	**B**	22	7 7 1 7 0 0	Pēteris Eriņš
LVA-2		4	3 1 0 0 0 0	Ēriks Gopaks
LVA-3	**HM**	12	7 3 1 1 0 0	Ieva Ozola
LVA-4	**HM**	7	7 0 0 0 0 0	Jānis Smilga
LVA-5		8	1 3 0 4 0 0	Jevgēnijs Vihrovs
LVA-6	**HM**	8	1 0 0 7 0 0	Normunds Vilciņš

Liechtenstein   LIE

LIE-1		2	0 1 0 1 0 0	Ricarda Gassner
LIE-2	**B**	19	6 4 0 6 3 0	Florian Meier

Lithuania   LTU

LTU-1	**S**	27	7 6 2 5 7 0	Vaidotas Juronis
LTU-2	**HM**	8	7 0 0 0 1 0	Paulius Kantautas
LTU-3	**HM**	7	7 0 0 0 0 0	Edvard Poliakov
LTU-4		2	1 1 0 0 0 0	Rolandas Glotnis

| LTU-5 | **B** | 20 | 7 2 0 5 6 0 | Greta Kuprijanovaitė |
| LTU-6 | **HM** | 13 | 7 5 0 1 0 0 | Linas Klimavičius |

Luxembourg   LUX

LUX-1	**B**	15	7 7 0 1 0 0	Pierre Haas
LUX-2		0	0 0 0 0 0 0	Jerome Urhausen
LUX-3	**B**	18	7 7 1 1 0 2	Jingran Lin
LUX-4		9	4 4 0 0 1 0	Philippe Schram
LUX-5	**B**	14	0 7 0 5 2 0	Marc Sinner
LUX-6	**HM**	9	7 0 0 1 1 0	Gregoire Genest

Macau   MAC

MAC-1	**B**	17	7 7 0 2 1 0	Chi Choi Wong
MAC-2		3	1 0 0 2 0 0	Hou Meng Ip
MAC-3		1	1 0 0 0 0 0	Iat Kei Chan
MAC-4	**HM**	10	2 7 0 0 1 0	Chi Tou Lam
MAC-5		9	1 6 0 0 2 0	Chao Keong Lo
MAC-6	**HM**	9	1 1 0 0 7 0	Kan Chun Leong

The former Yugoslav Republic of Macedonia   MKD

MKD-1	**S**	30	7 7 7 2 7 0	Bodan Arsovski
MKD-2	**HM**	10	7 3 0 0 0 0	Filip Talimdzioski
MKD-3	**B**	14	7 1 0 4 2 0	Predrag Gruevski
MKD-4	**B**	16	6 7 0 3 0 0	Bojan Joveski
MKD-5		6	1 5 0 0 0 0	Stefan Lozanovski
MKD-6	**B**	15	6 2 0 7 0 0	Stefan Stojchevski

Malaysia   MAS

| MAS-2 | **S** | 28 | 7 7 1 6 7 0 | Loke Zhi Kin |
| MAS-5 | | 3 | 0 1 0 2 0 0 | Muhammad Syafiq Johar |

Mauritania   MRT

MRT-1		2	1 1 0 0 0 0	Cheikh Tidjani Ahmed Vall
MRT-2		0	0 0 0 0 0 0	Ahmed Salem Sid'Ahmed
MRT-3		1	0 0 1 0 0 0	Sidi Mohamed Ahmed Salem
MRT-4		2	1 1 0 0 0 0	Mohamed Salem Abdellahi
MRT-5		1	0 0 1 0 0 0	Mouadh Ahmed Bah
MRT-6		2	0 1 0 1 0 0	Saadna Mohamed Vadel

Mexico   MEX

MEX-1	**B**	16	7 7 0 0 2 0	Erik Alejandro Gallegos Baños
MEX-2	**B**	16	7 7 1 1 0 0	Manuel Guillermo López Buenfil
MEX-3	**B**	16	3 7 0 6 0 0	César Bibiano Velasco
MEX-4	**HM**	13	7 3 0 3 0 0	Flavio Hernández González
MEX-5		7	4 3 0 0 0 0	Luis Ángel Isaías Castellanos
MEX-6		6	2 3 0 1 0 0	César Ernesto Rodríguez Angón

Republic of Moldova   MDA

MDA-1	**B**	15	7 7 0 1 0 0	Dragoş Gîlcă
MDA-2		5	2 2 1 0 0 0	Iulian Gramaţki
MDA-3	**B**	14	7 3 0 2 2 0	Alexandru Grigoroi
MDA-4	**B**	18	7 7 0 1 3 0	Andrei Iliaşenco
MDA-5		3	0 1 0 0 2 0	Mihai Indricean
MDA-6	**B**	19	7 7 0 5 0 0	Andrei Ivanov

Mongolia   MNG

MNG-1		6	5 0 0 1 0 0	Batkhuyag Batsaikhan
MNG-2	**B**	21	7 7 0 5 2 0	Enkhzaya Enkhtaivan
MNG-3	**B**	14	7 2 0 5 0 0	Luvsanbyamba Buyankhuu
MNG-4		7	4 0 0 3 0 0	Tsogbayar Idertsogt
MNG-5	**B**	17	7 7 1 0 2 0	Zorigoo Ochirkhuyag
MNG-6	**HM**	7	7 0 0 0 0 0	Byambadorj Otgonsuren

Montenegro   MNE

MNE-1		0	0 0 0 0 0 0	Sonja Vuksanovic
MNE-2		3	1 1 0 1 0 0	Milorad Vujkovic
MNE-3		7	1 5 1 0 0 0	Radovan Krtolica
MNE-4	**HM**	13	7 0 1 5 0 0	Nikola Milinkovic

Morocco   MAR

MAR-1		4	3 1 0 0 0 0	Naoufal Aghbal
MAR-2		3	0 0 0 3 0 0	Abdelatif Mharchi
MAR-3		7	2 3 0 1 1 0	Abdelilah El Hadfaoui
MAR-4		9	2 0 0 5 2 0	Rim Hariss
MAR-5		7	5 1 0 0 1 0	Amine Boutaybi
MAR-6		2	0 1 0 0 1 0	Mohammed Benslimane

Netherlands   NLD

NLD-1	S	24	7 7 1 2 7 0	Wouter Berkelmans
NLD-2	B	21	7 7 1 6 0 0	Raymond van Bommel
NLD-3	HM	9	7 0 0 1 1 0	Harm Campmans
NLD-4		9	4 3 0 2 0 0	Saskia Chambille
NLD-5	HM	11	7 4 0 0 0 0	David Kok
NLD-6		5	5 0 0 0 0 0	Maarten Roelofsma

New Zealand   NZL

NZL-1	B	16	7 1 0 6 2 0	Malcolm Granville
NZL-2	HM	8	7 1 0 0 0 0	Ben Kornfeld
NZL-3		7	3 2 0 2 0 0	Stephen Mackereth
NZL-4	HM	10	7 2 0 1 0 0	David Shin
NZL-5	HM	7	7 0 0 0 0 0	Ha Young Shin
NZL-6		5	5 0 0 0 0 0	Michael Wang

Nigeria   NGA

NGA-1		2	0 2 0 0 0 0	Isa Modibbo Ismail
NGA-2		2	0 0 0 2 0 0	Damilola Durajaiye
NGA-3		5	3 1 0 1 0 0	Ogunkola Opemipo Oladapo
NGA-4		0	0 0 0 0 0 0	Mark Donald Kanayochukwu Amobi
NGA-5		3	1 1 0 1 0 0	Jombo Eniweke Eunice
NGA-6		5	2 3 0 0 0 0	Puis Aje Onah

Norway   NOR

NOR-1	HM	9	7 1 0 1 0 0	Sivert Bocianowski
NOR-2	HM	9	7 1 1 0 0 0	Karl Erik Holter
NOR-3	B	19	7 5 0 0 7 0	Sondre Kvamme
NOR-4		7	6 1 0 0 0 0	Bernt Ivar Nødland
NOR-5		2	0 1 1 0 0 0	Felix Tadeus Prinz
NOR-6	B	14	7 7 0 0 0 0	Jarle Stavnes

Pakistan   PAK

PAK-1	B	14	7 1 1 3 2 0	Waqar Ali Syed
PAK-2		2	1 1 0 0 0 0	Isfar Tariq
PAK-3		1	0 1 0 0 0 0	Ahmad Bilal Aslam
PAK-5		3	0 2 0 1 0 0	Absar Ul Hassan
PAK-6		1	1 0 0 0 0 0	Qasim Mahmood

Panama   PAN

PAN-1   **HM** 12   7 3 0 1 1 0   Antonio Fan

Paraguay   PAR

PAR-1   5   0 4 0 1 0 0   Ariel Schvartzman Cohenca
PAR-2   4   3 1 0 0 0 0   Marcos Martínez Sugastti
PAR-3   5   4 0 0 1 0 0   Juan José Mongelós Wollmeister
PAR-4   0   0 0 0 0 0 0   Claudia Vanessa Montanía Portillo

Peru   PER

PER-1   **S** 25   5 7 1 4 7 1   César Cuenca Lucero
PER-2   **S** 28   6 7 1 7 7 0   Julián Alonso Mejía Cordero
PER-3   **B** 21   5 4 0 5 7 0   Percy Augusto Guerra Ríos
PER-4   **S** 26   6 7 1 5 7 0   Ricardo Jesús Ramos Castillo
PER-5   **B** 16   7 7 0 0 2 0   Raúl Arturo Chávez Sarmiento
PER-6   **S** 28   7 6 1 7 7 0   Tomás Miguel Angles Larico

Philippines   PHI

PHI-1   1   0 0 1 0 0 0   Carlo Francisco Echavez Adajar
PHI-2   2   1 1 0 0 0 0   Earl John Rosales Chua
PHI-3   **B** 15   7 4 0 2 2 0   Carmela Antoinette Sio Lao
PHI-4   8   1 2 0 5 0 0   Jonathan Santos Wong

Poland   POL

POL-1   **S** 29   7 7 7 1 7 0   Tomasz Kociumaka
POL-2   **B** 22   6 7 1 1 7 0   Karol Konaszyński
POL-3   **S** 29   7 7 1 7 7 0   Jakub Oćwieja
POL-4   **B** 22   7 1 0 7 7 0   Damian Orlef
POL-5   **B** 22   7 7 1 0 7 0   Tomasz Pawłowski
POL-6   **B** 16   7 0 1 1 7 0   Jakub Witaszek

Portugal   POR

POR-1   **HM** 9   7 1 0 1 0 0   Gonçalo Pereira Simões Matos
POR-2   **B** 15   7 1 1 6 0 0   João Morais Carreira Pereira
POR-3   **B** 23   7 7 0 6 3 0   Jorge Ricardo Landeira da Silva Miranda
POR-4   **S** 29   7 7 1 7 7 0   Pedro Manuel Passos de Sousa Vieira
POR-5   **HM** 8   7 1 0 0 0 0   Raúl Queiroz Do Vale de Noronha Penaguião
POR-6   **B** 15   7 5 0 1 2 0   Ricardo Correia Moreira

Puerto Rico   PRI

PRI-1		2	1 0 0 1 0 0	Aravind Arun
PRI-2	HM	12	7 3 0 2 0 0	George Arzeno
PRI-3		2	0 1 0 1 0 0	Kidhanis De Jesus
PRI-4		1	0 1 0 0 0 0	Jose Pacheco
PRI-5		3	0 1 0 2 0 0	Sara Rodriguez
PRI-6		3	2 0 0 1 0 0	Alan Wagner

Romania   ROU

ROU-1	G	34	7 7 7 6 7 0	Elena Mădălina Persu
ROU-2	S	27	7 7 2 4 7 0	Andrei Deneanu
ROU-3	S	24	7 7 2 6 2 0	Francisc Bozgan
ROU-4	B	23	4 7 1 5 6 0	Tudor Pădurariu
ROU-5	B	22	7 7 1 0 7 0	Marius Tiba
ROU-6	G	33	7 7 2 7 7 3	Omer Cerrahoglu

Russian Federation   RUS

RUS-1	G	32	7 4 7 7 7 0	Vladimir Bragin
RUS-2	G	35	7 7 7 7 7 0	Marsel Matdinov
RUS-3	G	32	7 7 1 7 7 3	Gleb Nenashev
RUS-4	G	39	7 7 7 7 7 4	Victor Omelyanenko
RUS-5	S	30	7 7 2 7 7 0	Kirill Savenkov
RUS-6	G	35	7 7 7 7 7 0	Konstantin Tyshchuk

El Salvador   SLV

SLV-1	6	2 3 0 1 0 0	Julio Cesar Ayala Menjivar
SLV-2	4	0 3 0 1 0 0	Jaime Antonio Bermudez Huezo
SLV-3	3	1 0 0 1 1 0	Nahomy Jhopselyn Hernández Cruz

Serbia   SRB

SRB-1	G	34	7 7 7 6 7 0	Teodor von Burg
SRB-2	S	27	7 7 1 5 7 0	Luka Milićević
SRB-3	S	27	7 7 2 7 4 0	Dušan Milijančević
SRB-4	B	23	7 7 0 6 3 0	Vukašin Stojisavljević
SRB-5	S	29	7 7 1 7 7 0	Mihajlo Cekić
SRB-6		13	6 5 0 2 0 0	Stefan Stojanović

Singapore   SGP

SGP-1	S	25	7 3 7 1 7 0	Jia-Han Chiam
SGP-2	B	15	7 7 1 0 0 0	Tian Wen Daniel Low
SGP-3	S	24	7 4 1 5 7 0	Yao Chen Ivan Loh
SGP-4	HM	13	7 1 1 3 1 0	Jia Hao Barry Tng
SGP-5	B	21	7 7 0 5 2 0	Ryan Jun Neng Chan
SGP-6	B	18	7 7 1 3 0 0	Jeck Lim

Slovakia   SVK

SVK-1	B	16	7 1 1 7 0 0	Martin Bachratý
SVK-2	HM	8	7 1 0 0 0 0	Peter Csiba
SVK-3		7	3 2 0 0 2 0	Eduard Eiben
SVK-4	B	20	7 7 1 3 2 0	Michal Hagara
SVK-5	HM	11	7 2 1 1 0 0	Filip Sládek
SVK-6	HM	11	7 3 1 0 0 0	Jakub Uhrík

Slovenia   SVN

SVN-1	B	14	7 3 0 2 2 0	Matej Aleksandrov
SVN-2	HM	11	7 3 0 1 0 0	Gregor Grasselli
SVN-3		5	2 3 0 0 0 0	Nik Jazbinšek
SVN-4	HM	11	7 3 0 1 0 0	Anja Komatar
SVN-5	HM	12	7 0 0 5 0 0	Matjaž Leonardis
SVN-6		5	1 4 0 0 0 0	Primož Pušnik

South Africa   SAF

SAF-1	HM	11	7 3 0 1 0 0	Francois Conradie
SAF-2	B	15	7 5 0 1 2 0	Henry Thackeray
SAF-3	B	23	7 7 7 0 2 0	Liam Baker
SAF-4	HM	12	7 3 1 0 1 0	Sean Wentzel
SAF-5	HM	12	7 4 0 0 1 0	Arlton Gilbert
SAF-6	HM	11	7 4 0 0 0 0	Greg Jackson

Spain   ESP

ESP-1	B	20	5 3 0 5 7 0	Moisés Herradón Cueto
ESP-2	B	15	7 3 0 3 2 0	Iván Geffner
ESP-3		5	2 1 0 2 0 0	Jaime Roquero
ESP-4	B	15	7 7 1 0 0 0	Glenier Lázaro Bello Burguet
ESP-5	B	15	7 1 0 0 7 0	Ander Lamaison
ESP-6		1	0 1 0 0 0 0	Alberto Merchante

Sri Lanka   LKA

LKA-1	**B**	22	7 7 0 6 2 0	Lajanugan Logeswaran
LKA-2	**B**	16	7 7 0 0 2 0	Buddhima Ruwanmini Gamlath Gamlath Ralalage
LKA-3	**HM**	7	0 0 0 7 0 0	Mahamarakkalage Dileepa Yasas Fernando
LKA-4		8	2 0 0 6 0 0	Kirshanthan Sundararajah
LKA-5	**HM**	9	7 0 0 2 0 0	Don Anton Tharindu Kumar Warnakula Warnakulaarachchiralalage
LKA-6	**HM**	12	0 7 0 5 0 0	Mukundadura Yasod Sankalpa Fonseka

Sweden   SWE

SWE-1	**HM**	12	7 3 0 2 0 0	Hampus Engsner
SWE-2	**B**	14	7 7 0 0 0 0	Gabriel Isheden
SWE-3	**HM**	8	7 0 0 1 0 0	Jenny Johansson
SWE-4	**HM**	7	7 0 0 0 0 0	Eric Larsson
SWE-5	**HM**	12	7 2 1 0 2 0	Rickard Norlander
SWE-6	**B**	17	7 1 0 2 7 0	Peter Zarén

Switzerland   SUI

SUI-1		10	5 2 0 1 2 0	Jürg Bachmann
SUI-2	**B**	14	6 1 0 5 2 0	Hrvoje Dujmovic
SUI-3	**B**	19	7 7 2 0 3 0	Eben Freeman
SUI-4	**HM**	10	7 2 0 0 1 0	Clemens Pohle
SUI-5	**B**	18	5 7 1 4 1 0	Raphael Steiner
SUI-6	**HM**	8	7 0 0 1 0 0	Pascal Su

Syria   SYR

SYR-1		1	0 1 0 0 0 0	Budour Khalil
SYR-2		1	0 1 0 0 0 0	Basel Ahmad
SYR-3		2	0 1 0 1 0 0	Nour Maamary
SYR-4		3	0 2 0 1 0 0	Mokhtar Nadren
SYR-5		0	0 0 0 0 0 0	Dina Ibrahim

Taiwan   TWN

TWN-1	**S**	28	7 7 0 7 7 0	Han Lin Hsieh
TWN-2	**S**	27	7 7 1 5 7 0	Yu-Fan Tung
TWN-3	**S**	28	7 7 1 6 7 0	Chien-Yi Wang
TWN-4	**G**	33	7 7 1 7 7 4	Hsin-Po Wang
TWN-5	**S**	24	7 3 0 7 7 0	Yi-Chan Wu
TWN-6	**S**	25	7 7 1 3 7 0	Chen-Yu Yang

Tajikistan   TJK

TJK-1		10	1 3 0 6 0 0	Olimjon Pirahmad
TJK-2	**B**	19	7 7 0 5 0 0	Muhamadjon Shoev
TJK-3	**S**	24	7 7 1 5 4 0	Inomzhon Mirzaev
TJK-4	**B**	14	2 7 0 5 0 0	Komron Giesiev
TJK-5		11	0 6 0 5 0 0	Shohin Sadullozoda
TJK-6		4	1 0 0 3 0 0	Khusravi Elmurod

Thailand   THA

THA-1	**S**	30	7 7 2 7 7 0	Nipun Pitimanaaree
THA-2	**S**	27	7 7 1 7 5 0	Pakawut Jiradilok
THA-3	**S**	31	7 7 5 5 7 0	Pongpak Bhumiwat
THA-4	**S**	29	7 7 1 7 7 0	Supanat Kamtue
THA-5	**G**	35	7 7 7 7 7 0	Suthee Ruangwises
THA-6	**S**	29	7 7 1 7 7 0	Thanard Kurutach

Trinidad and Tobago   TTO

TTO-1	**HM**	11	7 2 0 1 1 0	Chandresh Amrit Ramlagan
TTO-2	**HM**	10	1 7 0 2 0 0	Kerry Shastri Singh
TTO-3		3	3 0 0 0 0 0	Prithvi Ramakrishnan
TTO-4		2	1 1 0 0 0 0	Bjorn Varuun Ramroop
TTO-5		1	0 1 0 0 0 0	Amanda, Mary Aleong
TTO-6		1	0 1 0 0 0 0	Siddhartha Jahorie

Tunisia   TUN

TUN-1	**B**	14	7 2 0 5 0 0	Beyrem Khalfaoui
TUN-2		8	6 0 0 0 2 0	Amine Marrakchi
TUN-3		2	0 1 0 1 0 0	Imen Allouch
TUN-4		2	1 1 0 0 0 0	Mannai Nidhal
TUN-5		1	0 0 0 1 0 0	Youssef Khalfalli

Turkey   TUR

TUR-1	**S**	26	7 7 0 5 7 0	Vefa Göksel
TUR-2	**S**	29	7 7 1 7 7 0	Fehmi Emre Kadan
TUR-3	**S**	25	7 7 1 7 3 0	Ufuk Kanat
TUR-4	**G**	34	7 7 7 6 7 0	Süreyya Emre Kurt
TUR-5	**S**	28	7 7 7 0 7 0	Melih Üçer
TUR-6	**G**	35	7 7 7 7 7 0	Umut Varolgüneş

Turkmenistan   TKM

TKM-1	B	19	7 7 0 5 0 0	Nazar Emirov
TKM-2	B	21	7 6 0 5 3 0	Kakamyrat Tushiyev
TKM-3	B	20	7 7 0 4 2 0	Gaygysyz Hojanazarov
TKM-4		12	3 2 0 5 2 0	Dovlet Akyniyazov
TKM-5		1	1 0 0 0 0 0	Archyn Seyidov
TKM-6	S	24	7 7 1 7 2 0	Agageldi Samedov

Ukraine   UKR

UKR-1	B	21	7 7 0 0 7 0	Igor Kudla
UKR-2	S	24	7 7 1 7 2 0	Anastasiya Lysakevych
UKR-3	G	33	7 7 5 7 7 0	Vitaliy Senin
UKR-4	G	35	7 7 1 7 7 6	Nazar Serdyuk
UKR-5	B	14	1 7 0 5 1 0	Darya Shchedrina
UKR-6	G	35	7 7 7 7 7 0	Bogdan Veklych

United Arab Emirates   UAE

UAE-1		2	2 0 0 0 0 0	Alya Alqaydi
UAE-2		1	0 0 0 1 0 0	Fatmah Al Salami
UAE-3		0	0 0 0 0 0 0	Kaltham Yousef Tahnon
UAE-4		0	0 0 0 0 0 0	Meera Saeed Khalifa Almehairi
UAE-5		0	0 0 0 0 0 0	Alya Abdulla Musabbeh Salem Alqaydi

United Kingdom   UNK

UNK-1	B	20	7 7 1 4 1 0	Chris Bellin
UNK-2	S	31	7 7 7 7 3 0	Luke Betts
UNK-3	G	32	7 7 7 7 4 0	Tim Hennock
UNK-4	S	25	7 3 1 7 7 0	Peter Leach
UNK-5	B	21	7 7 0 0 7 0	Sean Moss
UNK-6	S	28	6 7 1 7 7 0	Preeyan Parmar

United States of America   USA

USA-1	G	35	7 7 7 7 7 0	John Berman
USA-2	S	25	7 7 1 3 7 0	Wenyu Cao
USA-3	G	34	7 7 3 7 7 3	Eric Larson
USA-4	S	29	7 7 1 7 7 0	Delong Meng
USA-5	S	30	7 7 7 2 7 0	Evan O'Dorney
USA-6	S	29	7 7 1 7 7 0	Qinxuan Pan

Uruguay   URY

URY-1		4	0 3 0 1 0 0	Germán Chiazzo
URY-2	HM	8	7 1 0 0 0 0	Matias Escuder
URY-3		0	0 0 0 0 0 0	Ari Najman
URY-4		3	1 1 0 1 0 0	Javier Peraza
URY-5		4	0 1 0 3 0 0	Ismael Valentín Rodríguez Brena
URY-6		2	0 2 0 0 0 0	Nicolas Uviedo

Uzbekistan   UZB

UZB-1	B	16	7 6 0 2 1 0	Abror Pirnapasov
UZB-2	B	16	3 7 0 6 0 0	Azizkhon Nazarov
UZB-3	S	24	7 7 1 6 3 0	Diyora Salimova
UZB-4		12	1 6 0 5 0 0	Gulomjon Abdurashitov
UZB-5	HM	12	0 7 0 5 0 0	Ibrokhimbek Akramov
UZB-6		5	0 0 0 5 0 0	Jasurbek Bahramov

Venezuela   VEN

VEN-1		10	3 1 0 6 0 0	Carmela Acevedo
VEN-2		3	2 1 0 0 0 0	Mauricio Marcano

Vietnam   VNM

VNM-1	B	16	6 7 0 3 0 0	Nguyễn Xuân Cương
VNM-2	G	39	7 7 7 7 7 4	Hà Khương Duy
VNM-3	S	25	7 3 1 7 7 0	Nguyễn Hoàng Hải
VNM-4	S	29	7 7 1 7 7 0	Phạm Hy Hiếu
VNM-5	G	33	7 7 7 7 5 0	Phạm Đức Hùng
VNM-6	B	19	2 7 0 6 4 0	Tạ Đức Thành

Zimbabwe   ZWE

ZWE-1		3	2 0 0 0 1 0	Mahdi Finnigan
ZWE-2		2	1 1 0 0 0 0	Oscar Takabvirwa

Chapter 7
Mathematical Guests of Honour

The highlight of the 2009 Olympiad was the celebration of the IMO's 50th anniversary. Six of the world's leading mathematicians, all of them highly successful former IMO participants, were invited as our guests of honour. Their presentations made the anniversary celebration a truly memorable and unique event. We are deeply grateful to:

7.1 Béla Bollobás

Cambridge, United Kingdom; Memphis, USA

Béla Bollobás (* 1943) was the first super star among the IMO participants. He participated in the three very first IMOs and won 2 gold medals and a bronze medal (see also the recollection of István Reiman). This record stood until it was broken by László Lovász and József Pelikán in the years 1963–1966. By the way, both are very well known in the scene and present at our 50th anniversary. Béla Bollobás has worked in various areas of mathematics, including functional analysis, combinatorics, graph theory and percolation.

He wrote his first doctorate in discrete geometry under the supervision of László Fejes Tóth and Paul Erdős in 1967. After spending a year in Oxford he went to Cambridge, where he received his Ph.D. in 1972. He has been a Fellow of Trinity College, Cambridge since 1970; in 1996 he was appointed to the Jabie Hardin Chair of Excellence at the University of Memphis, and in 2005 he was awarded a Senior Research Fellowship at Trinity College.

He proved numerous important results on extremal graph theory, functional analysis, the theory of random graphs, graph polynomials and percolation. For example, with Paul Erdös he proved sharp results about the structure of dense graphs; he proved that the chromatic number of the random graph on n vertices is asymptotically $n/2\log n$ and he proved basic discrete isoperimetric inequalities.

In addition to over 350 research papers on mathematics, he has written several books, including the research monographs "Extremal Graph Theory", "Random Graphs" and "Percolation" (with Oliver Riordan), the introductory text books "Modern Graph Theory", "Combinatorics" and "Linear Analysis", and the collection of problems "The Art of Mathematics — Coffee Time in Memphis", with drawings by his wife Gabriella Bollobás, an accomplished sculptor and painter.

Béla Bollobás has had 35 Ph.D. students, including 9 former IMO participants: József Balogh (IMO 1989–1990: S, S), Timothy Gowers (IMO 1981: G, Fields Medal in 1998 and one of our guests of honor), Ian Holyer (IMO 1972: B), Imre Leader (IMO 1981: S, leader of the IMO team of UNK for several years), Robert Morris (IMO 1998: B), Luke Pebody (IMO 1991–1993: S, S, H), Oliver Riordan (IMO 1988–1990: G, S, S), Amites Sarkar (IMO 1992: B) and Jerzy Wojciechowski (IMO 1976, 1978: B, T).

Béla Bollobás is a Foreign Member of the Hungarian Academy of Sciences; in 2007 he was awarded the Senior Whitehead Prize by the London Mathematical Society.

He is also a sportsman, having represented Oxford University at modern pentathlon and Cambridge University at fencing.

The Lion and the Christian, and Other Pursuit and Evasion Games

Béla Bollobás

Abstract.   In this note we shall show that a playful question in recreational mathematics can quickly lead to mathematical results and unsolved problems.

1 An Arena in Rome

1 An aged Lion and an agile young Christian in the arena.

Béla Bollobás

Department of Pure Mathematics and Mathematical Statistics, University of Cambridge, Cambridge CB3 0WB, UK; and Department of Mathematical Sciences, University of Memphis, Memphis TN 38152, USA. e-mail: B.Bollobas@dpmms.cam.ac.uk

It was a beautiful spring day in the second year of the reign of Marcus Ulpius Nerva Traianus, Emperor of Rome. The gladiatorial games put on to celebrate the Emperor's victory over Dacebalus, King of Dacia, and the conquest of his entire kingdom, had been in full swing for sixty days, and to the joy of the good citizens of Rome, showed no sign of abating. The amazing Colosseum that had been built a few years before was almost full, and the people were eagerly awaiting the treat this beautiful new day would bring them.

It all started well: the gladiators were magnificent — they were skilful and extremely brave, and many a fighter received his freedom from the Emperor. However, when it came to one of the favourite parts of the show, *men against beasts*, there was disappointment in the air: only one Lion and one Christian appeared in the arena. The Lion was big enough, but on a closer examination he turned out to be old, way past his prime, while the Christian was a fit young man. It soon transpired that the Lion and the Christian had the same top speed, so the good citizens of Rome began to wonder whether the Lion would ever catch the Christian.

While watching this pathetic spectacle, the more mathematical-minded citizens of Rome came up with the following problem they were confident they could solve within a few minutes.

The Lion and Christian Problem. *A Lion and a Christian (each considered to be a single point) move about in a closed circular disc with the same maximum speed. Can the Lion catch the Christian in finite time?*

As we shall see, this problem is not as easy as it looks: it is a prime example of the large family of *pursuit and evasion* problems. In this brief paper we shall discuss some aspects of these problems.

Note that in order to turn their conundrum into a mathematical problem, the citizens have made some simplifications usual in mathematics and physics: the Lion and the Christian have been turned into *points*, and the Colosseum has become a *circular* closed disc, rather than an oval, which would be a better approximation of the shape of the arena. The disc is assumed to be *closed*, so that the points corresponding to the Lion and Christian are allowed to be on the circle bounding the disc. Rather importantly, the catch is required to be in *finite time*. (Needless to say, it is assumed that at the start of the game the Lion and the Christian are not in the same spot.) The question is about a 'clever' Lion and a 'clever' Christian: the Lion 'plays' as well as possible, and so does the Christian.

Contrary to the story above, this question was not invented by the good citizens of Rome close to two thousand years ago, but in the 1930s by the German–British mathematician Richard Rado, who called it the *Lion and Man* problem. Its solution (the second solution below) remained the standard answer for about twenty years, when the Russian–British mathematician Abram S. Besicovitch, the Rouse Ball Professor in Cambridge, found a brilliant and unexpected solution (the third solution below). After this, the question and its answer by 'Bessie', as Besicovitch was known affectionately, were extensively popularized by the great British mathematician John E. Littlewood, Bessie's predecessor in the Rouse Ball chair, in his *Miscellany* [6] (see also [2]).

In the next section we sketch three solutions to this problem; the section after that is about various extensions of it, the fourth section is about some finer points of pursuit and evasion games, and the last section is about further results and open problems.

2 Solutions

In this section we shall give three solutions to the LC problem, coming to different conclusions.

First Solution: Curve of Pursuit. Clearly, the best strategy for the Lion is to run right towards the Christian. What should the Christian do? He should run round the perimeter of the disc at full speed. What develops then is that the Christian runs around the circle, and the Lion is running along the 'curve of pursuit'. Although this curve is not too easy to describe explicitly, even when the Lion starts at the centre of the disc and the Christian starts his run on the boundary, as in Figure 2, one can show that during this chase the Lion gets arbitrarily close to the Christian, *without ever catching him*.

Conclusion. *The Christian wins the LC game.*   □

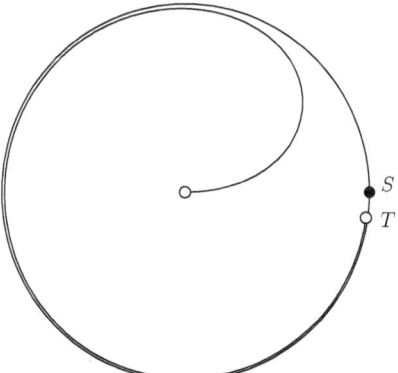

2 The curve of pursuit when the Lion starts from the centre and the Christian runs a full circle starting and ending at S. When the Christian gets back to S, the Lion is not far away, at T.

Curve of pursuit problems, like the one above, have been around in mathematics for almost three centuries, with the pursuer and the pursuit having different speeds. The standard formulation of these *pure pursuit* problems is that of a dog racing towards his master who is walking in a field. (See Puckette [8] and Nahin [7].)

To give two other solutions, we shall need some notation. First, we shall write B for the position of the 'Beast', the Lion, and C for that of the Christian, suppressing the dependence of these points on the time. We may and shall assume that the action takes place in the unit disc D with centre O, and the maximum speed of the contestants is 1. Also, we shall take D to be *closed*, i.e. we shall take it together with its boundary. (In fact, much of the time it will make no difference whether D is closed or not, although occasionally, like in the first solution we have just seen, and in the second solution below, it is convenient to take the boundary circle as part of our disc D.)

Second Solution: Stay on the Radius. The Lion decides to be not as greedy as in the first solution, but adopts the cunning plan of *staying on the segment OC* and, subject to this constraint, running towards the Christian at maximum speed. What happens if, as before, the Christian races along the boundary of D at full speed? Let us assume, for simplicity, that the Christian, C, starts at a point S of the big (radius 1) circle and runs in anticlockwise direction, and the Lion, B, starts at its centre, O.

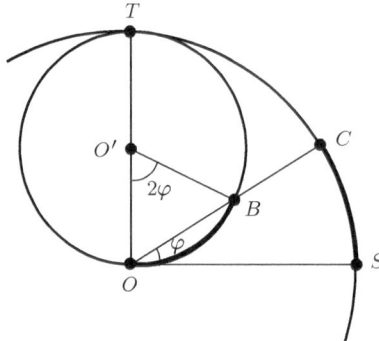

3 The Lion's path when he starts from the centre and the Christian from S. When the Christian, running along the big circle, reaches T, so does the Lion.

Claim. If the Lion follows his 'stay-on-the-radius' strategy, then he runs along the small (radius $1/2$) circle touching the line OS at O and the big circle at T, where the ST-arc of the big circle is a quarter-circle, as in Figure 3. Even more, wherever C is on the arc ST of the big circle, B is the intersection of the segment OC with the small circle. In particular, when the Christian gets to T, so does the Lion.

To justify this Claim, all we have to show is that the length of the SC-arc of the big circle is precisely the length of the OB-arc of the small circle (with centre O'). This is immediate from the two facts that a) the line OS is a tangent of the small circle, so the angle $BO'O$ is twice as large as the angle COS, and b) the radius of the big circle is twice the radius of the small one.

Conclusion. *The Lion wins the LC game.*   □

This solution shows that in the first argument we were too hasty to assume that the best strategy for the Lion is to run right towards the Christian. In fact, as we have seen, the Lion can do better by adopting a less obvious strategy: he tries to cut off the path of the Christian by running towards his future position.

Some readers may well suspect that we are trying to fool them, and the Christian can escape by simply reversing his direction along the boundary of the circle. This is not the case at all: as the Lion is on the radius leading to the Christian, for the Lion it makes no difference in which direction the Christian runs. By reversing his direction along the boundary the Christian gains nothing, since according to his strategy the Lion just stays on the radius, and so runs towards the same side as the Christian. No matter how often the Christian changes his direction while running at full speed along the boundary, he will be caught in exactly the same time as before. And slowing down just leads to a swifter end.

This is where the problem and its solution rested for about twenty years: the 'stay-on-the-radius' strategy is a quick win for the Lion. This is nice, but rather boring: for a mathematician it is not worth a second glance. Then, in the 1950s, a thunderbolt struck, when Besicovitch found the following beautiful argument. It is not clear what prompted Besicovitch to consider the problem again: it is quite possible that he wanted to mention it in an after-dinner talk to mathematics undergraduates in his College, Trinity.

Third Solution: Run along a Polygonal Path of Infinite Length. In this solution we describe a strategy for the Christian. Trivially, we may assume that in the starting position the Christian is in $C_1 \neq O$, the Beast is in $B_1 \neq C_1$ and the length $\overline{OC_1}$ is r_1, where $0 < r_1 < 1$.

Claim. Suppose that there are positive numbers $t_1, t_2, \ldots$ such that $\sum_i t_i$ is infinite but $\sum_i t_i^2 < 1 - r_1^2$. Then the Christian can escape.

To show this, split the (infinite) time into a sequence of intervals, of lengths $t_1, t_2, \ldots$. We shall 'review the situation' at times $s_i = \sum_{j=1}^{i-1} t_j$, $i = 1, 2, \ldots$, calling the time period between s_i and $s_{i+1} = s_i + t_i$ the i-th *step*. For convenience, set $t_0 = r_1$ so that $\sum_{i=0}^{\infty} t_i^2 < 1$.

Suppose that at time s_i the Christian is at point $C_i \neq O$, the Beast in $B_i \neq C_i$, and C_i is at a distance $r_i = \overline{OC_i}$ from the centre, where $r_i^2 = \sum_{j=0}^{i-1} t_j^2 < 1$. (Since the Christian starts at C_1, this is consistent with our earlier assumption. The condition $C_i \neq O$ is utterly unimportant: its only role is to make the description below slightly easier.) Let ℓ_i be the line through O and C_i. At the i-th step the Christian runs for time t_i in a straight line perpendicular to ℓ_i, in the direction that takes him away from B_i as much as possible. To be precise, if B_i is not on ℓ_i, but in the interior of one of the half-planes bounded by ℓ_i then B_i runs away from this half-plane, otherwise (i.e. if B_i is on ℓ_i) either direction will do. During this step the Christian runs away from the line ℓ_i; as the Lion starts either on ℓ_i or on its 'wrong' side, he has no chance of catching the Christian in this time step (see Figure 4). In particular, $C_{i+1} \neq B_{i+1}$ and $C_{i+1} \neq O$.

How far is the Christian from the centre at time s_{i+1}? By Pythagoras' theorem, the square of this distance $\overline{OC_{i+1}}$ is precisely $r_i^2 + t_i^2 = \sum_{j=0}^{i} t_j^2 = r_{i+1}^2 < 1$. Hence the polygonal path $C_1 C_2 \ldots$ of *infinite* length the Christian runs along remains in the

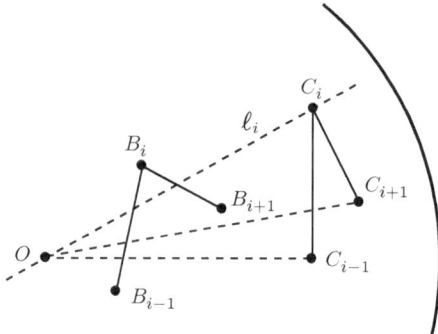

4 The polygonal path of the Christian.

disc D, and the Christian is not caught during this run, completing the proof of our Claim.

Finally, the Claim gives a winning strategy for the Christian, since it is easy to choose a sequence $t_1, t_2, \ldots$ satisfying the conditions in our Claim: e.g. we may take $t_i = 1/(i+r)$ for r large enough, since $\sum_{i=1}^{\infty} 1/i = \infty$ and $\sum_{i=1}^{\infty} 1/i^2$ is finite.

Conclusion. *The Christian wins the LC game.*   □

Clearly, this is 'where the buck stops'. Bessie's solution is indeed correct: using his polygonal path strategy, the Christian can indeed escape, no matter what the Lion does. The first 'solution' collapsed since we had no right to assume that the Lion rushes straight at the Christian; the second 'solution' collapsed since it was still based on the unjustified assumption that the best strategy for the Christian is to run along the boundary. The third, *correct* solution shows that, like a boxer on the ropes, in the second solution the Christian puts himself at a disadvantage by restricting his movements.

There is an obvious variant of the strategy in the third solution, which is slightly 'better' for the Christian: in the i-th step, he can run in a direction perpendicular to the B_iC_i line, starting in the direction which initially takes him closer to O. Unless the points B_i, C_i and O are collinear, this is better for the Christian in the sense that he stays further away from the boundary.

3 Variants

There are numerous variants of the LC game: here we shall mention only a few, leaving the exact formulations to the reader. Let us start with a question which must have occurred to the reader some time ago.

5 Birds catching a Fly.

1. Does the shape of the arena (within reason) matter? *Could the Lion win in an oval arena, the kind the Romans really had? Or in a triangular arena? Could the Lion drive the Christian into a corner of the triangle and then devour him?*

No doubt, a reader who has paid a little attention to the strategies we have given will see through this question in an instant.

Turning to less trivial variants, here are two results proved by Croft [4].

2. Birds catching a Fly. *There are some Birds and a Fly in the d-dimensional closed unit ball, each with the same maximum speed. What is the minimal number of Birds that can catch the Fly?*

What Bessie's solution of the LC game tells us is that for $d = 2$ one bird is not enough; it is easy to see that two birds suffice. In general, $d - 1$ birds do not suffice, but d do.

3. Uniformly bounded curvature. *If the Christian is forced to run along a curve of uniformly bounded curvature, then the Lion can win the LC game.*

Roughly, if the Christian cannot change direction arbitrarily fast, then the Lion can catch him.

What happens if we have many Lions and one Christian, but the game is played on the entire plane rather than in a bounded arena? The answer was given by Rado and Rado [9] and Janković [5].

4. Many Lions in the Plane. *Finitely many Lions can catch a Christian in the plane in finite time if and only if the Christian is in the interior of the convex hull of the Lions.*

Finally, here is a problem which is still open.

5. Two Lions on a golf course. *Can two Lions catch the Christian on a golf course with finitely many rectifiable lakes?*

Needless to say, the assumptions are that neither the Christian nor the Lions are allowed to step into the lakes, and the boundaries of the lakes are 'nice' in a technical sense (see Figure 6).

6 Lions trying to catch a Christian.

4 Mathematical Formalism

Having read three 'solutions' of the LC game, the reader is entitled to wonder whether we actually know what a 'winning strategy' really means. If the game is played using alternate moves, then there is no problem: in every time step, knowing the position of the game, the player has to decide what to do next. In particular, a winning strategy amounts to a choice of moves ending in a win, no matter what the opponent does. However, if the game is played in continuous time, we have to be more careful. Indeed, can we say precisely what a strategy is in the continuous LC game? To answer this question precisely, we need some definitions. We shall write $|x|$ for the modulus of a number x and also for the length of a vector x. In particular, if $x, y \in D$ then $|x - y|$ denotes the distance between the points x and y.

Suppose the Lion starts at x_0 and the Christian at y_0, and both have maximum speed 1. A *Lion path* is a map f from $[0, \infty)$ to the unit disc D such that $f(0) = x_0$ and $|f(t) - f(t')| \leq |t - t'|$ for all times $t, t' \geq 0$. (Such a map f is said to be 'Lipschitz', with constant 1.) Similarly for a Christian path. When the Lion follows a path f then at time t he is at $f(t)$, and when the Christian follows a path g then at time t he is at $g(t)$.

Let $\mathscr{B}$ be the set of Lion paths (the paths of the beast) and $\mathscr{C}$ the set of the Christian paths. Then a *strategy* for the Christian is a map Φ from $\mathscr{B}$ into $\mathscr{C}$ such that if $f_1, f_2 \in \mathscr{B}$ agree up to time t_0 (i.e. $f_1(t) = f_2(t)$ for all $0 \leq t \leq t_0$) then $\Phi(f_1)$ and $\Phi(f_2)$ agree on $[0, t_0]$. This 'no lookahead' condition tells us that $\Phi(f)(t)$ depends only on the restriction of f to the interval $[0, t]$. A Lion strategy $\Psi : \mathscr{C} \to \mathscr{B}$ is defined similarly. A Christian strategy Φ is *winning* if $\Phi(f)(t) \neq f(t)$ for every path $f \in \mathcal{B}$ and for every $t \geq 0$. And a Lion strategy Ψ is *winning* if for every path $g \in \mathcal{C}$ of the Christian there is a time $t \geq 0$ such that $\Psi(g)(t) = g(t)$.

Are these the 'correct' definitions? A moment's thought tells us that they are, since we want to allow strategies *without delay*, like the 'curve of pursuit' and 'stay on the radius' strategies of the Lion in the first two 'solutions' above.

Note that these definitions make sense in more general circumstances. For instance, the arena need not be the disc: any set in the plane or 3-dimensional space would do. In fact, so would any metric space (a space with a sensible 'distance' function). Consequently, pursuit–evasion games like the LC game make good sense in these general metric spaces. As we wish to give the Lion a chance to catch the Christian, we shall always assume that the playing field (our metric space) contains a path from the Lion to the Christian.

Having defined what we mean by a winning an LC-type pursuit–evasion game, let us ask a question our readers may find surprising. We know that in the disc Bessie's strategy is a win for the Christian;

but could it happen that the Lion also has a winning strategy?

Surely many readers would agree that this question is not only surprising but also downright crazy. How on earth could both have a winning strategy? Of course not, we say loud and clear:

if both had winning strategies then with each playing his own winning strategy we would get a game which is a win for the Christian and also a win for the Lion — a blatant contradiction.

A little thought should tell us that, once again, we have been too hasty: this 'argument' is sheer nonsense. Indeed, how can *both* play their winning strategies? Suppose that the outcome of a game in which both play their winning strategies, Φ and Ψ, is a Lion path f and a Christian path g. Then $\Phi(f) = g$ and $\Psi(g) = f$; in particular, $\Psi(\Phi(f)) = f$, i.e. f is a fixed point of the composite map $\Psi \circ \Phi$ mapping $\mathcal{B}$ into itself. *But why should the composite map $\Psi \circ \Phi$ have a fixed point at all?* There is no reason why it should.

Thus, in a general pursuit–evasion game, there are two basic questions. Does the Lion have a winning strategy? Does the Christian have a winning strategy? Can all four conceivable combinations of answers occur?

If the moves in our game alternate, say, if the moves take place at times 1, 2, etc., with the Christian moving at odd times and the Lion at even times, then there is no problem with playing a strategy against another. Thus in this case it cannot happen that both players have winning strategies. What gives the present continuous problem an entirely different complexion is that whoever *plays his strategy* can react *instantaneously* to the move (really 'path' or 'trajectory') of the other. This is certainly the right definition if we want to allow strategies like the Lion's 'stay on the radius' strategy, in which the Lion is shadowing the move of the Christian, instantly reacting to any change of speed or direction of his prey.

There are also several other natural questions. For example, are there 'nice' winning strategies? The most obvious way a strategy can be 'nice' is that it is *continuous*: it maps 'nearby' paths into 'nearby' paths. (Formally, a Christian strategy $\Phi : \mathcal{B} \to \mathcal{C}$ is continuous if for every $f_0 \in \mathcal{B}$ and $\varepsilon > 0$ there is a $\delta > 0$ such that if $f_1 \in \mathscr{B}$ is such that $|f_0(t) - f_1(t)| < \delta$ for every t then $|\Phi(f_0)(t) - \Phi(f_1)(t)| < \varepsilon$ for every t. And similarly for a Lion strategy.)

Also, what happens if we play the *bounded-time game*, i.e. the entire game must run its course by a fixed time T? In this version the Christian wins if he can stay alive up to time T. And what happens if we postulate that the playing field is 'nice'?

If everything goes as we 'feel' it should, then our mathematical formalism is rather wasted. However, the results in the final section show that this is far from the case: there are several unexpected twists.

5 Results and Open Problems

In this final section we shall give a selection of results from a recent paper by Bollobás, Leader and Walters [3], and state some open problems.

We shall play the bounded-time game on a playing field which is just about as nice as possible: a *compact* metric space, so that every sequence has a subsequence converging to a point of the space, like a closed interval, a closed disc or a closed ball.

What about the following two statements?

1. At least one of the players has a winning strategy.
2. At most one of the players has a winning strategy.

Certainly, both statements feel true. In fact, by considering the discretised version of the game, one can show that the first statement is true in the best of all possible worlds: in a compact metric space, at least one of the players has a winning strategy for the bounded-time game. It turns out that not even this assertion is trivial to prove; for a proof, see [3].

And what about the second statement, which feels just as much true? Surprisingly, this statement is false.

A game in which both players have winning strategies. *Let the playing field be the closed solid cylinder*

$$D \times I = \{(a,z) : a \in D \text{ and } 0 \le z \le 1\},$$

with the distance of two points $(a,z), (b,u) \in D \times I$ defined to be

$$\max\{|a-b|, |z-u|\}.$$

At the start, C is at the centre of the top of the cylinder (a unit disc), and B is at the centre of the bottom. Then both players have winning strategies.

Proof. Having defined this game, it is very easy to justify these assertions. Indeed, the Christian can win if he can make a small move which takes him away from being exactly above the Lion, since from then on he can simply run the Bessie strategy, ignoring the height, and so survives for ever. But can he get away from above the Lion in time $t_0 < 1/2$, say? Unexpectedly, this can be done very easily. We encourage the reader to find a simple argument for this.

And how can the Lion win? That is even easier. He keeps his disc coordinate the same as the Christian, and increases his height with speed 1 until he catches him. This will happen by time 1. Note that the Lion makes use of the fact that the distance on the solid cylinder is not the usual Euclidean distance, but the so-called ℓ_∞ *distance*, the maximum of the distances in the disc D and the interval I.   □

Note that the simple strategies in the proof above are examples of strategies which cannot be played against each other.

Let us return to the original Christian and Lion game in the closed unit disc D. Bessie's strategy is a winning strategy for the Christian, and it turns out that it can be discretised to show that the Lion does *not* have a winning strategy. But is Bessie's strategy continuous? By considering the positions in which O, B and C are collinear, we can see that it is not. In fact, considerably more is true.

Continuity in the original Lion and Christian game. *In the original game, neither player has a continuous winning strategy.*

Furthermore, with the Lion starting in the origin, for any continuous strategy of the Christian, there is a Lion path catching the Christian by time 1.

Proof. As we know that the Lion does not have a winning strategy, we have to prove only the second assertion. For this, we need a classical result from topology, *Brouwer's fixed point theorem,* stating that every continuous map $\varphi : D \to D$ has a fixed point, i. e. a point $x \in D$ such that $\varphi(x) = x$ (see, e.g., [1], p. 216).

Let then $\Phi : \mathscr{B} \to \mathscr{C}$ be a continuous Christian strategy. For every $z \in D$, let h_z be the constant speed straight path from 0 to z, reaching z at time 1, i.e. set $h_z(t) = tz$ (assuming, as we do, that the origin is the centre of D.) Then $z \mapsto h_z$ is a continuous map from D into $\mathscr{B}$. Consequently, $z \mapsto \Phi(h_z)(1)$ is a continuous function mapping D into D. By Brouwer's fixed point theorem, this map has a fixed point $z_0 \in D$: $\Phi(h_{z_0})(1) = z_0$. Hence, if the Lion follows h_{z_0}, and the Christian plays using his strategy Φ then the Lion catches the Christian at time 1.   □

Rather curious phenomena can occur if the arena is not compact, even if otherwise it is as nice as possible. For example, Alexander Scott noted that in the LC game played on the *open* interval $(0, 1)$ *both* the Lion and the Christian have winning strategies. Indeed, suppose the Lion starts at $2/3$ and the Christian at $1/3$. Then

$$f(t) \mapsto \Phi(f)(t) = f(t)/2$$

is a winning strategy for the Christian, and

$$g(t) \mapsto \Psi(g)(t) = \max\{2/3 - t, g(t)\}$$

is a winning strategy for the Lion.

However, as we mentioned earlier, it was shown in [3] that a bounded-time LC game played on a compact field cannot be too pathological: at least one of the players has a winning strategy.

In a bounded-time LC game played on a compact field at least one of the players has a winning strategy.

Finally, let us leave the reader with two open questions which arise naturally from this result.

1. Is there a bounded-time LC game in which neither player has a winning strategy?

2. Is there an unbounded-time LC game played on a compact field in which neither player has a winning strategy?

For other results and questions, the reader is referred to the original paper [3].

Acknowledgement. I am grateful to Gabriella Bollobás for her pen drawings.

References

1. Béla Bollobás, *Linear analysis: an introductory course. Second edition.* Cambridge University Press, Cambridge, 1999, xii+240 pp.
2. Béla Bollobás, *The art of Mathematics — coffee time in Memphis.* Cambridge University Press, New York, 2006, xvi+359 pp.
3. Béla Bollobás, Imre Leader, and Mark Walters, *Lion and Man — can both win?* Preprint, September 14, 2009, 24 pages; http://arxiv.org/abs/0909.2524 .
4. Hallard T. Croft, *"Lion and Man": a postscript.* Journal of the London Mathematical Society **39** (1964), 385–390.
5. Vladimir Janković, *About a man and lions.* Matematički Vesnik **2** (1978), 359–361.
6. John E. Littlewood, *Littlewood's Miscellany. Edited and with a foreword by B. Bollobás.* Cambridge University Press, Cambridge, 1986, vi + 200 pp.
7. Paul J. Nahin, *Chases and Escapes —The Mathematics of Pursuit and Evasion.* Princeton University Press, Princeton/NJ, 2007, xvi + 253 pp.
8. C.C. Puckette, *The curve of pursuit.* Mathematical Gazette **37** (1953), 256–260.
9. P.A. Rado and R. Rado, Mathematical Spectrum **7**(1974/75), 89–93.

7.2 Timothy Gowers

Cambridge, United Kingdom

Timothy Gowers (* 1963) participated once at an IMO, 1981 in the United States of America, and came home with a perfect score of 42 points. He studied Mathematics at Cambridge University, where he obtained his PhD degree in 1990. His PhD advisor was Béla Bollobás, himself a successful former IMO participant and guest of honor at the IMO anniversary celebration. After an appointment at University College London from 1991–1995, he moved to the University of Cambridge as Professor of Mathematics and fellow of Trinity College.

Timothy Gowers was awarded the Prize of the European Mathematical Society in 1996, and the Fields Medal in 1998. He was elected Fellow of the Royal Society in 1999.

The research of Timothy Gowers relates two areas of mathematics that at seem rather unrelated, functional analysis and combinatorics. Combinatorics usually stands for discrete structures with isolated points, while functional analysis deals with huge, usually infinite-dimensional spaces that are important in Mathematics as well as Theoretical Physics. Few mathematical areas seem further apart than these two. What these do have in common, it is said, is that they pose problems that are relatively easy to formulate but apparently rather difficult to prove. Among the best known achievements of Timothy Gowers are those that establish fruitful relations between these two fields.

Functional analysis was founded in the 1920's and 1930's by a group of mathematicians around Stefan Banach in Lviv, then Poland. They would often meet to discuss mathematics in a famous café called "Scottish Café", and described many of their conjectures in a notebook that is now known as the "Scottish Book"; the kind of spaces they considered became later known as "Banach spaces". A number of conjectures from those pioneering days had to wait until they were solved by Timothy Gowers, using tools from combinatorics. For instance, he constructed a Banach space with almost no symmetry: this is of particular importance to Physicists as well because much of Physics is described in terms of Banach spaces, and symmetries are among the most fundamental concepts in Physics. Gowers' Banach space has provided a counterexample to several other conjectures in functional analysis.

Timothy Gowers has been quite active in works to popularize Mathematics, for instance as the author of the book *Mathematics: a very short introduction*. He co-edited the *Princeton Companion to Mathematics*, an extensive overview of Mathematics involving well-known contributors. He started a *web blog* partly in order to supplement this book after it appeared, inspired by the example of another pioneer of "21st century ways of doing, communicating, and teaching mathematics",

Terence Tao. He is exploring further innovative ways of teaching mathematics. For instance, just recently he started a new project called *polymath*, with the intention to produce mathematics collaboratively using the comment functionality of his web blog.

How do IMO Problems Compare with Research Problems?

Timothy Gowers

1 Introduction

Many people have wondered to what extent success at the International Mathematical Olympiad is a good predictor of success as a research mathematician. This is a fascinating question: some stars of the IMO have gone on to extremely successful research careers, while others have eventually left mathematics (often going on to great success in other fields). Perhaps the best one can say is that the ability to do well in IMO competitions correlates well with the ability to do well in research, but not perfectly. This is not surprising, since the two activities have important similarities and important differences.

The main similarity is obvious: in both cases, one is trying to solve a mathematical problem. In this article, I would like to focus more on the differences, by looking at an area of mathematics, Ramsey theory, that has been a source both of Olympiad problems and of important research problems. I hope to demonstrate that there is a fairly continuous path from one to the other, but that the two ends of this path look quite different.

Many expositions of Ramsey theory begin by mentioning the following problem.

Problem 1.1. *There are six people in a room, and any two of them are either good friends or bitter enemies. Prove that there must either be three people such that any two of them are good friends, or three people such that any two of them are bitter enemies.*

If you have not seen this problem (though I would imagine that most IMO contestants have encountered it), then you should solve it before reading on. It is not hard, but one learns a lot from working out the solution for oneself.

Timothy Gowers
Department of Pure Mathematics and Mathematical Statistics, University of Cambridge, Cambridge CB3 0WB, UK. e-mail: W.T.Gowers@dpmms.cam.ac.uk

It is convenient to reformulate the problem before solving it, by stripping it of its irrelevant non-mathematical part (that is, its talk of people, friendship and enmity) and looking just at the abstract heart of the problem. One way of doing this is to represent the people by *points* in a diagram, and joining each pair of points by a (not necessarily straight) line. This gives us an object known as the *complete graph of order 6*. To represent friendship and enmity, we then colour these lines red if they join two people who are good friends and blue if they join two people who are bitter enemies. So now we have six points, with each pair of points joined by either a red line or a blue line. The standard terminology of graph theory is to call the points *vertices* and the lines *edges*. (These words are chosen because an important class of graphs is obtained by taking a polyhedron and forming a graph out of its vertices and edges. In such an example, there will be pairs of points not joined by edges, unless the polygon is a tetrahedron: these are therefore *incomplete* graphs.) Our task is now to prove that there must be a red triangle or a blue triangle, where a *triangle* in this context means a set of three edges that join three vertices.

To prove this, pick any vertex. It is joined by edges to five other vertices, so by the pigeonhole principle at least three of those edges have the same colour. Without loss of generality, this colour is red. There are therefore three vertices that are joined by red edges to the first vertex. If any two of these three vertices are joined by a red edge, then we have a red triangle. If not, then all three pairs of vertices from that group of three must be joined by blue edges, which gives us a blue triangle. QED

Let us define $R(k,l)$ to be the smallest number n such that if you colour each edge of the complete graph of order n red or blue, then you must be able to find k vertices such that any two of them are joined by a red edge, or l vertices such that any two of them are joined by a blue edge. We have just shown that $R(3,3) \leq 6$. (If you have not already done so, you should find a way of colouring the edges of the complete graph with five vertices red or blue in such a way that you do *not* get a red triangle or a blue triangle.)

A simple generalization of the argument we used to prove that $R(3,3) \leq 6$ can also be used to prove the following result, which is due to Erdős and Szekeres.

Theorem 1.2. *For every k and l, we have the inequality*

$$R(k,l) \leq R(k-1,l) + R(k,l-1).$$

Once again, if you have not seen this before, then you should prove it for yourself. (It is much easier than a typical IMO problem.) And you can then prove, by an easy inductive argument, that the inequality above implies that $R(k,l) \leq \binom{k+l-2}{k-1}$. (The main addtional step is to note that $R(k,1) = 1$, or, if that bothers you, then the slightly safer $R(k,2) = k$, which allows you to get the induction started.)

This tells us that $R(3,4) \leq 10$. However, the true answer is in fact 9. Proving this is a more interesting problem – not too hard, but it involves an extra idea. From that and the inequality of Erdős and Szekeres, we may deduce that $R(4,4) \leq R(3,4) + R(4,3) = 18$, which turns out to be the correct answer: to show this you need to think of a red-blue colouring of the edges of the complete graph of order 17 such that no four vertices are all joined by red edges and no four vertices are all joined by blue edges. Such a graph exists, and it is rather beautiful: as ever, I would not want to spoil the fun by saying what it is.

We do not have to go much further than this before we enter the realms of the unknown. Using the Erdős-Szekeres inequality again we find that $R(3,5) \leq R(2,5) + R(3,4) = 5 + 9 = 14$, which turns out to be the actual value, and then that $R(4,5) \leq R(4,4) + R(3,5) = 32$. In 1995, McKay and Radziszowski showed, with a great deal of help from a computer, that in fact $R(4,5) = 25$. The best that is currently known about $R(5,5)$ is that it lies between 43 and 49.

It is not clear that the correct value of $R(5,5)$ will ever be known. Even if the answer is 43, a brute-force search on a computer through all of the $2^{\binom{43}{2}}$ red-blue colourings of the complete graph of order 43 would take far too long to be feasible. Obviously, there are ways of cutting this search down, but so far not by enough to make the computation feasible. At any rate, even if somebody does eventually manage to calculate $R(5,5)$, it is highly unlikely that $R(6,6)$ will ever be known. (It is known to be between 102 and 165.)

Why, you might ask, do we not try to find a *theoretical* argument rather than an ugly argument that checks huge numbers of graphs on a computer? The reason is that the largest graphs that avoid k vertices that are all joined by red edges or l vertices that are all joined by blue edges tend to be rather unstructured. In this respect, the graphs that demonstrate that $R(3,3) > 5$, $R(3,4) > 8$ and $R(4,4) > 17$ are rather misleading, since they have plenty of structure. This seems to be an example of the so-called "law of small numbers". (For a simpler example of this, take the fact that the first three primes, 2, 3 and 5, are consecutive Fibonacci numbers. This fact is of no significance whatsoever: there just aren't that many small numbers around so one expects coincidences.)

We therefore find ourselves in the unsatisfactory situation that there is probably no theoretical argument that gives an exact formula for $R(k,l)$, and therefore the best one can do is try to find clever search methods on a computer when k and l are small. This may sound a bit defeatist, but Gödel has taught us that we cannot just assume that everything we want to know has a proof. In the case of small Ramsey numbers, we do not learn anything directly from Gödel's theorem, since we could in principle calculate them by brute force, even if not in practice. However, the general message that nice facts do not have to have nice proofs still applies, and has an impact on the life of a research mathematician, which can be summed up in the following general problem-solving strategy, which I do not recommend to participants in Mathematical Olympiads.

Strategy 1.3. *When you are stuck on a problem, sometimes the best thing to do is give up.*

As a matter of fact, I do not entirely recommend it to research mathematicians either, unless it is coupled with the following rather more positive principle, which again I do not recommend to participants in Mathematical Olympiads.

Strategy 1.4. *If you cannot answer the question, then change it.*

2 Asymptotics of Ramsey Numbers

One of the commonest ways of changing a mathematical question when we find ourselves in a situation such as the one just described, faced with a quantity that we do not think we can calculate exactly, is to look for the best approximations that we can find, or at least to prove that the quantity must lie between L and U, where we try to make L (called a *lower bound*) and U (called an *upper bound*) as close as we can.

We have already obtained an upper bound for $R(k,l)$: the bound in question was $\binom{k+l-2}{k-1}$. For simplicity let us look at the case where $k = l$. Then we obtain the upper bound $\binom{2(k-1)}{k-1}$. Can we match that with a comparable lower bound?

Before we try to answer this question, we should first think about roughly how large $\binom{2(k-1)}{k-1}$ is. A fairly good approximation (but by no means the best known) is given by the formula $(k\pi)^{-1/2}4^{k-1}$, which we can think of as growing at about the same speed as 4^k (since the ratios of successive values of this function get closer and closer to 4).

This is a pretty large function of k. Is there any hope of finding a lower bound of anything like that size?

If by "finding" you mean writing down a rule that tells you when to colour an edge red and when to colour it blue, then the answer is that to find an exponentially large lower bound is a formidably difficult unsolved problem (though there are some fascinating results in this direction). However, in 1947, Erdős came up with a simple but revolutionary method of obtaining an exponentially large lower bound that does not involve finding one in this sense. Rather than give Erdős's proof, I shall just give the idea of the proof. It will be useful to introduce the following piece of terminology. If we have a red-blue colouring of the edges of the complete graph on n vertices, then let us call a set of vertices *monochromatic* if any two vertices in the set are joined by edges of the same colour.

Idea of Proof. *Do not attempt to find a colouring that works. Instead, choose the colours randomly and prove that the average number of monochromatic sets of size k is less than 1.*

If we can do that, then there must be a graph with no monochromatic sets of size k, since otherwise the average would have to be at least 1. The calculations needed to make this argument work turn out to be surprisingly simple, and they show that $R(k,k)$ is at least $\sqrt{2}^k$. (In fact, they give a slightly larger estimate than this, but not by enough to affect this discussion.)

The good news is that this lower bound is exponentially large. The bad news is that $\sqrt{2}^k$ is a *lot* smaller than 4^k. Can one improve one or other of these bounds? This is a central open problem in combinatorics.

Problem 2.1. *Does there exist a constant $\alpha > \sqrt{2}$ such that for all sufficiently large k we have the lower bound $R(k,k) \geq \alpha^k$, or a constant $\beta < 4$ such that for all sufficiently large k we have the upper bound $R(k,k) \leq \beta^k$?*

A more ambitious question is the following.

Problem 2.2. *Does the quantity* $R(k,k)^{1/k}$ *tend to a limit, and if so what is that limit?*

Probably $R(k,k)^{1/k}$ does tend to a limit. There are three natural candidates for what the limit might be: $\sqrt{2}$, 2 and 4. I have seen no truly convincing argument in favour of one of these over the other two.

There has been only a tiny amount of progress on these problems for several decades. So should we give up on them too? Definitely not. There is a profound difference between these extremely hard problems and the extremely hard problem of evaluating $R(6,6)$, which is that here one *expects* there to be a beautiful theoretical argument: it is just very hard to find. To give up the search merely because it is hard would be to go completely against the spirit of mathematical research. (Sometimes a single mathematician is well-advised to give up on a problem after spending a long time on it and getting nowhere. But here I am talking about a collective effort: pretty well all combinatorialists have at some time or another tried to improve the bounds for $R(k,k)$ and I am saying that this should continue until somebody eventually cracks it.)

3 What, in General, is Ramsey Theory?

A typical theorem in Ramsey theory concerns a structure that has many substructures that are similar to the main structure. It then says that if you colour the elements in the main structure with two colours (or more generally with r colours for some positive integer r), then you must be able to find a substructure all of whose elements have the same colour. For example, with Ramsey's theorem itself in the case where $k = l$, the structure is the complete graph of order $R(k,k)$ (or to be precise, the edges of the complete graph) and the substructures are all complete subgraphs of order k.

Here is another example, a famous theorem of van der Waerden.

Theorem 3.1. *Let r and k be positive integers. Then there exists a positive integer n such that if you colour the numbers in any arithmetic progression X of length n with r colours, then you must be able to find some arithmetic progression Y inside X of length k such that all the numbers in Y have been given the same colour.*

I could talk a great deal about van der Waerden's theorem and its ramifications, but that would illustrate less well some of the more general points I want to make about IMO problems and research problems. Instead, I want to move in a different direction.

4 An Infinitary Structure and an Associated Ramsey Theorem

Up to now, the structures we have coloured — complete graphs and arithmetic progressions — have been finite. There is a version of Ramsey's theorem that holds for
infinite complete graphs (another interesting exercise is to formulate this for yourself and prove it), but I want to look instead at a more complicated structure: the
space of all infinite 01-sequences that are "eventually zero". An example of such a
sequence is

00100111011000000000000000000000000000000000000000000000000000........

If s and t are two such sequences, and if the last 1 in s comes before the first 1 in
t, then we write $s < t$. (One can think of this as saying that all the action in s has
finished by the time the action in t starts.) If this is the case, then $s + t$ is another
01-sequence that is eventually zero. For example, you can add

00100111011000000000000000000000000000000000000000000000000000........

to

00000000000000011000110001100000000000000000000000000000000000........

and you will get the sequence

00100111011000011000110001100000000000000000000000000000000000........

Now let us suppose that we have sequences $s_1 < s_2 < s_3 < s_4 < \dots$. That is,
each s_i is a sequence of 0s and 1s, and all the 1s in s_{i+1} come after all the 1s in
s_i. (Note that $(s_1, s_2, s_3, \dots)$ is a sequence of sequences.) Therefore, if we take any
sum of finitely many distinct sequences s_i, then we will obtain another sequence
that belongs to our space. For example, we could take the sum $s_1 + s_2$, or the sum
$s_3 + s_5 + s_6 + s_{201}$. The set of all possible sums of this kind is called the *subspace
generated by* $s_1, s_2, s_3, \dots$.

Now the entire space of sequences that we are talking about can be thought of
as the subspace generated by the sequences $1000000\dots$, $0100000\dots$, $0010000\dots$,
$0001000\dots$, and so on. Thus, the structure of the entire space is more or less identi-
cal to that of any of its subspaces. This makes it an ideal candidate for a Ramsey-type
theorem. We can even guess what this theorem should say.

Theorem 4.1. *Let the 01-sequences that are eventually zero be coloured with two
colours. Then there must be an infinite collection $s_1 < s_2 < s_3 < \dots$ of sequences
such that all the sequences in the subspace generated by the s_i have the same colour.*

That is, however you colour the sequences, you can find a sequence of sequences
s_i such that s_1, s_2, $s_1 + s_2$, s_3, $s_1 + s_3$, $s_2 + s_3$, $s_1 + s_2 + s_3$, s_4 etc. all have the same
colour.

This theorem is due to Hindman, and is too difficult to be thought of as an exercise. However, one thing that *is* a simple exercise is to show that once you have Hindman's theorem for two colours then you can deduce the same theorem for any (finite) number of colours.

Hindman's theorem is usually stated in the following equivalent form, which is easier to grasp, but which relates less well to what I want to talk about in a moment. Proving the equivalence is another exercise that is not too hard.

Theorem 4.2. *Let the positive integers be coloured with two colours. Then it is possible to find positive integers $n_1 < n_2 < n_3 < \ldots$ such that all sums of finitely many of the n_i have the same colour.*

This version of the theorem concerns addition. What if we try to introduce multiplication into the picture as well? Almost instantly we are back in the world of the unknown, since even the following innocent looking question is an unsolved problem.

Problem 4.3. *Let the positive integers be coloured with finitely many colours. Is it always possible to find integers n and m such that n, m, $n + m$ and nm all have the same colour? Is it even possible to ensure merely that $m + n$ and mn have the same colour (except in the trivial case $m = n = 2$)?*

This looks very like an IMO problem. The difference is that it just happens to be far far harder (and one does not have the helpful knowledge that somebody has solved it and deemed it suitable for a mathematics competition).

5 From Combinatorics to Infinite-dimensional Geometry

We represent three-dimensional space by means of coordinates. Once we have done this, it is easy to define d-dimensional space for any positive integer d. All we have to do is express our concepts in terms of coordinates and then increase the number of coordinates. For example, a four-dimensional cube could be defined as the set of points (x, y, z, w) such that each of x, y, z and w is between 0 and 1.

If we want to (which we often do when we are doing university-level mathematics), we can even extend our concepts to *infinite*-dimensional space. For instance, an infinite-dimensional sphere of radius 1 can be defined as the set of all sequences $(a_1, a_2, a_3, \ldots)$ of real numbers that satisfy the condition $a_1^2 + a_2^2 + a_3^2 + \cdots = 1$. (Here I am using the word "sphere" to mean the surface of a ball rather than a solid ball.)

In our infinite-dimensional world, we also like to talk about lines, planes, and higher-dimensional "hyperplanes". In particular, we are interested in infinite-dimensional hyperplanes. How do we define these? Well, a plane going through the origin in three-dimensional space can be defined by taking two points $\mathbf{x} = (x_1, x_2, x_3)$ and $\mathbf{y} = (y_1, y_2, y_3)$ and forming all combinations $\lambda \mathbf{x} + \mu \mathbf{y}$ of these two points. (Here,

$\lambda\mathbf{x} + \mu\mathbf{y}$, when written out in coordinates, is $(\lambda x_1 + \mu y_1, \lambda x_2 + \mu y_2, \lambda x_3 + \mu y_3)$.)
We can do something similar in infinite-dimensional space. We take a sequence of
points $\mathbf{p}_1, \mathbf{p}_2, \mathbf{p}_3, \ldots$ (here each $\mathbf{p}_i$ will itself be an infinite sequence of real numbers)
and we take all combinations (subject to certain technical conditions) of the form
$\lambda_1\mathbf{p}_1 + \lambda_2\mathbf{p}_2 + \lambda_3\mathbf{p}_3 + \ldots$.

It turns out that if we look at the intersection of an infinite-dimensional sphere
with an infinite-dimensional hyperplane, then we get another infinite-dimensional
sphere. (Apart from the fact that all the dimensions are infinite, this is a bit like the
fact that if you intersect a sphere with a plane then you get a circle.) Let us call this
a *subsphere* of the original sphere. Once again we seem to be ideally placed for a
Ramsey-type theorem, since we have a structure (a sphere) with many substructures
(subspheres) that look exactly like the structure itself. Suppose that we colour an
infinite-dimensional sphere with two colours. Can we always find a subsphere that
has been coloured with only one colour?

There is some reason to expect a result like this to be true. After all, it is
quite similar to Hindman's theorem, in that both statements involve colouring some
infinite-dimensional object, defined by coordinates, and looking for a monochromatic infinite-dimensional subobject of a similar type. It is just that in Hindman's
theorem all the coordinates have to be 0 or 1.

Unfortunately, however, the answer to our new question is no. If $\mathbf{p}$ belongs to a
subsphere, then $-\mathbf{p}$ must belong to the same subsphere. So we could colour $\mathbf{p}$ red
if its first non-zero coordinate is positive and blue if its first non-zero coordinate is
negative. In that case, $\mathbf{p}$ and $-\mathbf{p}$ will always receive different colours. (Since the
squares of their coordinates have to add up to 1 they cannot all be zero.)

This annoying observation highlights another difference between IMO problems
and the kinds of questions that come up in mathematical research.

Principle 5.1. *Most conjectures that come up naturally in one's research are either
easy or wrong. One has to be lucky to stumble on an interesting problem.*

However, under these circumstances we can apply a variant of a strategy I mentioned earlier.

Strategy 5.2. *If the question you are thinking about turns out to be uninteresting,
then change it.*

Here is a small modification to the problem about colouring spheres, which turns
it from a bad problem into a wonderful one. Let us call a subsphere *c-monochromatic*
if there is a colour such that every point in the subsphere is within a distance c from
some point of that colour. We think of c as small, so what this is basically saying
is that we do not ask for all points in the subsphere to be red (say), but merely that
every point in the subsphere is *close* to a red point.

Problem 5.3. *If the infinite-dimensional sphere is coloured with two colours, and
c is a positive real number, then is it always possible to find a c-monochromatic
infinite-dimensional subsphere?*

This problem remained open for a long time and became a central question in the theory of Banach spaces, which are a formalization of the idea of infinite-dimensional space and one of the central concepts in research-level mathematics. Unfortunately, it too had a negative answer, but the counterexample that shows it is *much* more interesting and *much* less obvious than the counterexample to the bad version of the problem. It was discovered by Odell and Schlumprecht.

The example of Odell and Schlumprecht killed off the hope of a Hindman-like theorem for Banach spaces (except for one particular space where the similarity to the space of 01-sequences is more pronounced, for which I obtained such a theorem). However, it did not entirely destroy the connections between Ramsey theory and Banach-space theory, as we shall see in the next section.

Before we finish this section, let me mention another difference between IMO problems and research problems.

Principle 5.4. *A research problem can change from being completely out of reach to being a realistic target.*

To somebody with experience only of IMO problems, this may seem strange: how can the difficulty of a problem change over time? But if you look back at your own mathematical experience, you will know of many examples of problems that "became easy". For example, consider the problem of finding the positive real number x such that $x^{1/x}$ is the biggest it can be. If you know the right tools, then you argue as follows. The logarithm of $x^{1/x}$ is $\log x/x$, and the logarithm function is increasing, so the problem is equivalent to maximizing $\log x/x$. Differentiating gives us $(1 - \log x)/x^2$, which is zero only when $x = e$, and decreasing there. Hence, the maximum is at $x = e$.

That solution is fairly straightforward, both to understand and to find in the first place, but only if one knows a bit of calculus. So the problem is out of reach to people who do not know calculus and a realistic target to those who do. Something similar to this happens in mathematical research, but the additional point I am making is that it can be a *collective* phenomenon and not just an individual one. That is, there are many problems that are out of reach simply because the right technique has not been invented yet.

You might object that that does not really mean that the problem is out of reach: it just means that part of the work of solving it is to invent the right technique. In a way that is true, but it overlooks the fact that mathematical techniques are very often used to solve problems that were not the problems that originally motivated the technique. (For instance, Newton and Leibniz did not invent calculus so that we could maximize the function $x^{1/x}$.) Thus, it may well happen that Problem B becomes a realistic target because somebody has invented the right technique while thinking about Problem A.

I mention all this here because Odell and Schlumprecht built their counterexample by modifying (in a very clever way) an example that Schlumprecht had built a few years earlier for a completely different reason.

6 A Little Bit More About Banach Spaces

I am aware that I have not explained very clearly what a Banach space is, and I may have given the impression that the only notion of distance in infinite-dimensional spaces is what you get by generalizing Pythagoras's theorem and defining the distance of a point $(a_1, a_2, a_3, \dots)$ from the origin to be $\sqrt{a_1^2 + a_2^2 + a_3^2 + \dots}$.

However, other notions of distance are possible and useful. For instance, for any $p \geq 1$ we can define the distance from $(a_1, a_2, a_3, \dots)$ to the origin to be the p-th root of $|a_1|^p + |a_2|^p + |a_3|^p + \dots$. Of course, there will be sequences for which this number is infinite. We regard these sequences as not belonging to the space.

It is not obvious in advance that this will be a sensible notion of distance, but it turns out to have some very good properties. Writing **a** and **b** for the sequences $(a_1, a_2, a_3, \dots)$ and $(b_1, b_2, b_3, \dots)$, and writing $\|\mathbf{a}\|$ and $\|\mathbf{b}\|$ for the distances from **a** and **b** to the origin, usually known as the *norms* of **a** and **b**, we can express these properties as follows.

   (i) $\|\mathbf{a}\| = 0$ if and only if $\mathbf{a} = (0, 0, 0, \dots)$.
   (ii) $\|\lambda \mathbf{a}\| = |\lambda| \cdot \|\mathbf{a}\|$ for every **a**.
   (iii) $\|\mathbf{a} + \mathbf{b}\| \leq \|\mathbf{a}\| + \|\mathbf{b}\|$ for every **a** and **b**.

All three of these properties are properties that we are familiar with from the usual notion of distance in space. (Note that we can define the distance from **a** to **b** to be $\|\mathbf{a} - \mathbf{b}\|$.) A *Banach sequence space* is a set of sequences together with some norm that satisfies properties (i)–(iii) above, together with a more technical condition (called *completeness*) that I shall not discuss.

The particular example where $\|\mathbf{a}\|$ is defined to be $\left(\sum_{n=1}^{\infty} a_n^2 \right)^{1/2}$ is a very special kind of Banach space called a *Hilbert space*. I will not say what a Hilbert space is, except to say that it has particularly good symmetry properties. One of these good properties is that every subspace of a Hilbert space is basically just like the space itself. We have already seen this: when we intersected an infinite-dimensional sphere with an infinite-dimensional hyperplane we obtained another infinite-dimensional sphere. This property, that all subspaces are "isomorphic" to the whole space, does not seem to hold for any other space, so Banach himself asked the following question in the 1930s.

Problem 6.1. *Is every space that is isomorphic to all its (infinite-dimensional) subspaces isomorphic to a Hilbert space?*

To put that more loosely, is a Hilbert space the only space with this particularly good property? The difficulty of the question is that there are many ways that two infinite-dimensional spaces can be isomorphic, so ruling them all out for a non-Hilbert space and some carefully chosen subspace is likely to be hard.

This is another example of a problem that turned from impossible to possible as a result of developments connected with other problems, and I was lucky enough to be in the right mathematical place at the right time, so to speak. Some work of Komorowski and Tomczak-Jaegermann (incidentally, I am mentioning several mathematicians whose names will mean very little to most readers of this article, but decided against prefacing every single one with "a mathematician called") showed that if there was a counterexample to the problem, it would have to be rather nasty in a certain sense.

Now it is far from obvious that there could be a space as nasty as what would be required, but it so happened that a couple of years earlier Maurey and I had constructed just such a nasty space, and our nasty space was *so* nasty that for entirely different reasons it had no chance of being a counterexample to Banach's question. This raised the possibility that the answer to Banach's question was yes, because nice examples couldn't work, and nasty examples couldn't work either. In order to make an approach like this work, I found myself needing to prove a statement of the following kind.

Statement 6.2. *Every infinite-dimensional Banach space has an infinite-dimensional subspace such that all its subspaces are nice or all its subspaces are nasty.*

Now this has strong overtones of Ramsey theory: we could think of nice subspaces as "red" and nasty subspaces as "blue".

7 A Weak Ramsey-type Theorem for Subspaces

There is, however, one important difference between Statement 6.2 and our earlier Ramsey-theoretic statements, which is that here the objects we are colouring are (infinite-dimensional) *subspaces* rather than *points*. (However, I should point out that in Ramsey's theorem itself we coloured edges rather than vertices, so the idea of colouring something other than points is not completely new.) How do we fit this into our general framework?

It is in fact not too hard. The structures we are colouring can be thought of as "the structure of all subspaces of a given space". If we take any subspace, then all *its* subspaces form a structure of a similar kind to the structure we started with, so we can think of trying to prove a Ramsey-type theorem.

The best we could hope to prove would be something like this: if you colour all the subspaces of some space red or blue, then there must be a subspace such that all of *its* subspaces have the same colour. However, not too surprisingly, this turns out to be far too much to hope for, for both boring and interesting reasons. The boring reason is similar to the reason that we could not colour the points of an infinite-dimensional sphere and hope for a monochromatic infinite-dimensional subsphere. The interesting reason is that even if we modify the statement so that we are looking for subspaces that are *close* to all being the same colour (in some suitable sense of "close"), the results of Odell and Schlumprecht, which concerned colouring points, can be used fairly easily to show that we will not necessarily find them.

We appear to have reached a dead end, but in fact we have not, because for the application I had in mind, I did not need the full strength of a Ramsey theorem.

Instead, I was able to get away with a "weak Ramsey theorem", which I shall now briefly describe.

To do so, I need to introduce a curious-looking game. Suppose that we are given a collection Σ of sequences of the form $(\mathbf{a}_1, \mathbf{a}_2, \mathbf{a}_3, \dots)$, where all $\mathbf{a}_i$ are points in a Banach space. (It is important, here and in many previous places in this article, to keep in mind what the objects are that I am talking about. This can get quite complicated: Σ is a collection of sequences, as I have just said; but the terms in each sequence are themselves points in a Banach space, so they are sequences of real numbers, which is why I have written them in bold face. Thus, Σ is a set of sequences of sequences of real numbers. One could take this even further and say that each real number is represented by an infinite decimal, so Σ is a set of sequences of sequences of sequences of numbers between 0 and 9. But it is probably easier to think of the terms $\mathbf{a}_n$ as points in an infinite-dimensional space and forget about the fact that they have coordinates.) Given the collection Σ, Players A and B then play as follows. Player A chooses a subspace S_1. Player B then chooses a point $\mathbf{a}_1$ from S_1. Player A now chooses a subspace S_2 (which does not have to be a subspace of S_1) and player B chooses a point $\mathbf{a}_2$ from S_2. And so on. At the end of this infinite process, player B will have chosen a sequence $(\mathbf{a}_1, \mathbf{a}_2, \mathbf{a}_3, \dots)$. If this sequence is one of the sequences in the collection Σ, then B wins, and otherwise A wins.

Now obviously who wins this game depends heavily on what Σ is. For example, if there happens to be a subspace S such that it is impossible to find points $\mathbf{a}_n$ in S that form a sequence $(\mathbf{a}_1, \mathbf{a}_2, \mathbf{a}_3, \dots)$ in Σ, then A has the easy winning strategy of choosing S every single time, but if Σ contains almost all sequences then B will be expected to have a winning strategy.

Here, then, is the weak Ramsey theorem that turned out to be enough to prove a suitably precise version of Statement 6.2 and hence answer Banach's question (Problem 6.1). I have slightly oversimplified the statement.

Theorem 7.1. *In the above game, there will always be a subspace S such that either none of the sequences made out of points in S belong to Σ or B has a winning strategy if A only ever chooses subspaces of S.*

To see how one might call this a weak Ramsey theorem, let us colour a sequence red if it belongs to Σ and blue otherwise. Then the theorem says that we can find a subspace S such that either all the sequences built out of points in S are blue, or there are so many red sequences built out of points in S that if the game is confined to S then B has a winning strategy for producing red sequences.

In other words, we have replaced "all sequences in S are red" by "so many sequences in S are red that B has a winning strategy for producing them".

It is one thing to formulate such a statement and observe that it is sufficient for one's purposes, but quite another to prove it. This brings me to another difference between IMO problems and research problems, which is that the following problem-solving strategy is far more central to research problems than to IMO problems.

Strategy 7.2. *If you are trying to prove a mathematical statement, then search for a similar statement that has already been proved, and try to modify the proof appropriately.*

I would not want to say that this always works in research or that it never works in an IMO problem, but with IMO problems it is much more common to have to start from scratch.

Going back to the weak Ramsey theorem, it turned out to resemble another infinitary Ramsey theorem, due to Galvin and Prikry. The resemblance was close enough that I was able to modify the argument and prove what I needed. And luckily I had been to a course in Cambridge a few years earlier in which Béla Bollobás had covered the theorem of Galvin and Prikry.

8 Conclusion

I do not have much to say in conclusion that I have not already said. However, there is one further point that is worth making. If you are an IMO participant reading this, it may seem to you that your talent at solving Olympiad problems has developed almost without your having to do anything: some people are just good at mathematics. But if you have any ambition to be a research mathematician, then sooner or later you will need to take account of the following two principles.

Principle 8.1. *If you can solve a mathematical research problem in a few hours, then it probably wasn't a very interesting problem.*

Principle 8.2. *Success in mathematical research depends heavily on hard work.*

Even from the examples I have just given, it is clear why. When one sets out to solve a genuinely interesting research problems, one usually has only a rather hazy idea of where to start. To get from that hazy idea to a clear plan of attack takes time, especially given that most clear plans of attack have to be abandoned anyway – for the simple reason that they do not work. But you also need to be ready to spot the connections and similarities to other problems, and to have developed your own personal toolbox of techniques, bits of mathematical knowledge, and so on. Behind any successful research mathematician will be thousands of hours spent pondering mathematics, only very few of which will have directly led to breakthroughs. It is strange, in a way, that anybody is prepared to put in those hours. Perhaps it is because of a further principle such as this.

Principle 8.3. *If you are truly interested in mathematics, then hard mathematical work does not feel like a chore: it is what you want to do.*

7.3 László Lovász

Budapest, Hungary

László Lovász (* 1948) is one of the best known mathematicians in IMOs as well as in Mathematics, mainly in Discrete Mathematics and Theoretical Computer Science. In the years 1963–1966 he won 3 gold and 1 silver medals at the IMOs and was awarded two special prizes. He and József Pelikán, who reached exactly the same results, were the most successful IMO participants of those years. Even today, with more than 12346 students at all IMOs only 5 were more successful!

László Lovász is married and has 4 children. It's worth mentioning that also his son László Miklós Lovász won a gold and a silver medal in the last two years.

László Lovász is a very active researcher. He published already 300 research papers and several well known books. Among his many deep results let us mention the Perfect Graph Theorem (1972), the Lovász local lemma (1975) and Lenstra-Lenstra-Lovász lattice basis reduction (LLL) (1982).

He wrote books that are not only of interest to researchers but are

also used by IMO participants, like the celebrated book "Combinatorial problems and exercises" that became a standard for everyone who wants to learn combinatorics. In a similar spirit is the book "Discrete Mathematics" by him, József Pelikán and Lovász' wife, Katalin Vesztergombi, who also holds a Ph.D. in mathematics.

László Lovász earned his PhD already in 1970. In the very early age of 31 years he became corresponding member and with 37 year registered member of the Hungarian Academy of Sciences. Lovász is elected member of the European Academy of Sciences, Arts and Humanities, the Academia Europaea, the Nordrhein-Westfälische Akademie der Wissenschaften, the Deutsche Akademie der Naturforscher Leopoldina, the Russian Academy of Sciences, 2006 and the Royal Dutch Academy of Science.

In 1978 László Lovász became Professor of Geometry at the university in Szeged, in 1983 Professor of Computer Science at the university in Budapest, in 1993 Professor of Computer Science at Yale University. From 1999–2006 he was Senior Researcher at Microsoft Research, before he returned to Hungary, where he is currently the Director of the Mathematical Institute of the Eötvös Loránd University in Budapest.

László Lovász was honoured with several honorary degrees and positions. He is Doctor Honoris Causa of the universities of Waterloo/Canada, Szeged/Hungary, Budapest University of Technology and the University of Calgary/Canada. Moreover, among the many awards he received are: George Pólya Prize (1979), Ray D. Fulkerson Prize (1982), National Order of Merit of Hungary, (1998), Bolzano Medal

(1998), Wolf Prize, Israel, (1999), Knuth Prize (1999) and the Gödel Prize (2001). Currently, László Lovász is President of the International Mathematical Union: the association that organizes the International Congresses of Mathematicians (every four years) at which the Fields Medals and Nevanlinna Prizes are awarded.

Graph Theory Over 45 Years

László Lovász

Abstract. From 1963 to 1966, when I participated in IMOs, graph theory did not appear in the problem sets. In recent years, however, graph-theoretic problems were often given. What accounts for this? What is the role of graph theory in mathematics today? I will try to answer these questions by describing some of the many connections between graph theory and other areas of mathematics as I encountered them.

1 Introduction

Graph theory is not a new subject. The first result in graph theory was the solution of the Königsberg Bridges Problem in 1736 by Leonhard Euler, one of the greatest mathematicians of all times.

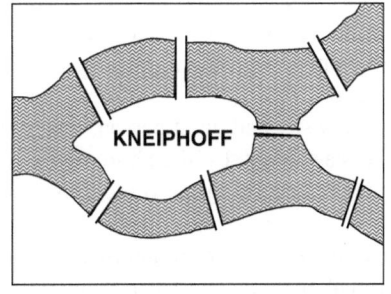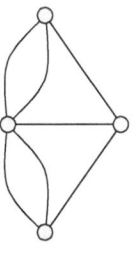

1 The bridges of Königsberg in Euler's time, and the graph modeling them.

László Lovász
Institute of Mathematics, Eötvös Loránd University, Budapest, Hungary.
e-mail: lovasz@cs.elte.hu

It started with a recreational challenge made by the citizens of Königsberg. The city was divided into four districts by the river Pregel (Figure 1), which were connected by seven bridges. This gave rise to a question: *Is it possible to walk in such a way that each bridge is crossed exactly once?*

Euler proved that such a walk is impossible. If we represent each district by a vertex and draw an edge between two vertices if there is a bridge that connects the corresponding districts, we get the little graph on the right hand side of Figure 1. The argument of Euler (which is simple and so well known that I will not reproduce it here) can be translated to this graph by examining the degrees of the vertices (that is, the number of edges incident with each vertex).

Many important results in graph theory were obtained in the nineteenth century, mostly in connection with electrical networks. However nobody thought to consider graph theory as an area of mathematics in its own right until the first book on graph theory, Dénes König's *Theorie der endlichen und unendlichen Graphen*, was published in 1936.

König taught in Budapest, and he had two very prominent students, Paul Erdős and Tibor Gallai. Many other Hungarian mathematicians of that generation got interested in graph theory and proved results which are considered fundamental today. Turán and Hajós are two examples.

I was introduced to graph theory, and thereby to mathematical research, quite early in life. My high school classmate and friend Lajos Pósa (himself a gold medalist at the IMO) met Erdős when he was quite young. Erdős gave him graph-theoretic problems to solve, and he was successful. When Pósa started high school, he wrote a paper with Erdős and then wrote more papers by himself. Later, when we met, he told me about other problems due to Erdős, and since I was able to solve one or two, Erdős and I were introduced. Erdős gave me unsolved problems, and thereafter I thought up some of my own, and so began my lifelong commitment to graph theory.

Many of you have probably heard of Erdős. He was not only one of the greatest mathematicians of the twentieth century but a special soul, who did not want to settle, did not want to have property (so that it would not distract him from doing mathematics), and traveled all the time. He was always surrounded by a big group of young people, and he shared with them new problems, research ideas, and results about which he learned during his travels.

Gallai was just the opposite — a very shy and quiet person, who preferred long, one-on-one conversations. When I was a student, I visited him regularly and learned a lot about graph theory and his ideas concerning the directions in which it was developing.

In those days, graph theory was quite isolated from mainstream mathematics. It was often regarded as recreational mathematics, and I was often advised by older mathematicians to do something more serious. I have worked in other areas of mathematics since then (algorithms, geometry, and optimization), but in one way or another there was always some graph-theoretic problem in the background.

Things have changed. Graph theory has become important over the course of the last decades, both through its applications and through its close links with other parts of mathematics. Let me describe some developments.

2 Discrete Optimization

I wrote my thesis under the guidance of Gallai on the problem of factors of graphs (today we call them matchings). The basic question is: *Given a graph, can we pair up its vertices so that the two vertices in each pair are adjacent?* (Such a pairing is called a *perfect matching*.) More generally, what is the maximum number of edges that are pairwise disjoint (that is, without a common endpoint)? A special case is that of bipartite graphs, i.e. graphs in which the vertices are divided into two classes, so that each edge connects two vertices in different classes. In this case, the answer was given by König in 1931: *The maximum number of pairwise disjoint edges in a bipartite graph is equal to the minimum number of vertices covering all edges.* Earlier, in 1914, König proved the following related theorem about bipartite graphs: *The minimum number of colors needed to color the edges of a graph so that edges with a common endpoint are colored differently is equal to the maximum degree of its vertices.*

Tutte extended the characterization of bipartite graphs with perfect matchings to all graphs (the condition is beautiful but a bit too complicated to state here). Many other matching problems remained unsolved, however (allowing me to write a thesis by solving some), and matching theory is still an important source of difficult, but not hopelessly difficult, graph-theoretic problems.

The Austrian mathematician Menger studied the following question in the 1920s: *Given a graph and two vertices s and t, what is the maximum number of mutually disjoint paths from s to t (meaning disjoint except for their common endpoints s and t)?* He proved that this number is equal to the minimum number of vertices (different from s and t) whose removal destroys all paths from s to t. This is a very useful identity, but Menger's proof does not tell us how to compute the aforementioned value. In fact it took thirty years until the American mathematicians Ford and Fulkerson defined flows in networks and used them to give an efficient algorithm for computing a maximum family of mutually disjoint s-t paths in a graph.

The matching problem and the disjoint path problem mentioned above are examples of optimization problems that are quite different from those one studies in calculus. There are many other graph-theoretic optimization problems, some of which, like the Traveling Salesman Problem, became very well known. In a typical optimization problem in analysis, we want to find the minimum or maximum of a function, where the function is "smooth" (differentiable) and defined on an interval. Many of you probably learned how to do this: we find the zeroes of the derivative of the function, and compare the values of the function at these points as well as at the ends of the interval. In discrete optimization, the situation is quite different: we want to optimize functions that are defined on a finite but large and complicated set (like the set of all matchings in a graph, where the function is the number of edges in the matching). These functions have no derivatives, and classical methods of analysis are useless.

There are several methods to attack such problems; perhaps the most successful proceeds by linear programming, which can be thought of as the art of solving systems of linear inequalities. Most of you could probably solve a system of two linear *equations* with two unknowns, or three linear equations with three unknowns. (You could, for example, eliminate one variable by subtracting the equations from each other, and then repeat.) Solving systems of linear *inequalities* is substantially more complicated, but also possible. The methods are analogous, but quite a bit more involved.

It is perhaps interesting to note that this algebraic problem can be translated to geometry by constructing a convex polyhedron (in a high-dimensional space) and reducing the optimization problem to optimizing a linear function over this polyhedron.

One important source of combinatorial optimization problems are hypergraphs. In an ordinary graph, every edge has two endpoints. We can generalize this, and allow edges that have any number of endpoints. Tibor Gallai called my attention to the fact that any problem in graph theory could be extended (usually in more than one way) to hypergraphs, and that virtually all of these hypergraph problems were unsolved (many of them still are).

For example, König's two theorems mentioned above remain perfectly meaningful for hypergraphs. The question is: how can we define "bipartite" hypergraphs so that the theorems remain not only meaningful but true? A first attempt is to assume that the vertices can be partitioned into two classes, so that every edge meets both classes (it could now contain more than one vertex from a class). We call such hypergraphs 2-*colorable*; they are interesting and important, but in this case, there are easy examples showing that both of König's theorems fail to hold for them. One can try other variations on the notion of bipartiteness, but none of these seem to work. In one of my first papers I managed to prove that the two theorems remain equivalent (even though there is no simple criterion under which they hold). This was a hypergraph-theoretic reformulation of a graph-theoretic conjecture of Berge, and the proof showed that hypergraph theory is useful not only for finding new research problems, but also for solving old problems.

3 Computer Science

Coming back to matching theory, many of us tried to obtain an analogue of Tutte's aforementioned characterization of graphs having a perfect matching for Hamilton cycles: *Given a graph, is there a cycle in it that goes through each vertex exactly once?* The problem is quite similar to the matching problem. It is also quite similar to the Euler cycle problem which started this paper, and which has an easy answer. My advisor Tibor Gallai and many of us were wondering why it was so much more difficult than the other two.

This time (around 1970) was also a time of rapid developments in computer science, in particular of the theory of algorithms and their complexity. In 1972–73, I spent a year in the US and learned about the newly developed theory of polynomial time algorithms and NP-complete problems. Polynomial time problems were "easy", or at least efficiently solvable. NP-complete problems are hard in the sense that every other problem in a rather wide class of problems can be reduced to them. It is widely expected that this is a real distinction, but no mathematical proof exists. This fundamental problem of complexity theory, usually stated as "P $\neq$ NP ?", was included in 2000 among the 7 most important open problems in mathematics.)

The theory of the complexity of algorithms thrilled me because it explained the difference between the matching problem and the Hamilton cycle problem: the first one was in P, and the other one was NP-complete!

When I returned to Hungary, I met a friend of mine, Péter Gács, who had spent a year in Moscow. Interrupting one another, we began to describe the great ideas we had learned about: Leonid Levin's work in Moscow, and the work of Cook and Karp in the US. As it turned out, they were independent developments of the same theory. (For about two weeks we thought we had a proof of the P $\neq$ NP conjecture. Nowadays, we would be more suspicious of simplistic ideas concerning a famous problem....)

Graph theory has become one of the most prominent areas of the mathematical foundation of computer science. We have seen that graph-theoretic problems motivated the "P = NP ?" problem and many more of the most interesting questions that arose in the development of complexity theory.

There is also an important connection that points in the other direction: to describe the process of a complicated computation in mathematical terms, one needs the notion of a directed graph. Steps in the computation are represented by vertices (often called "gates"), and an edge indicates that the output of one step is the input of the other. We can assume that these outputs are just bits, and the gates themselves can be very simple (it suffices to use just one kind, a NAND gate, which outputs TRUE if and only if at least one of the inputs is FALSE). All of the complexity of the computation goes into the structure of the graph.

I am sorry to report that we graph theorists have not achieved much in this direction. For example, the famous "P = NP ?" problem boils down to the following question: we want to design a network that can find out whether an arbitrary graph with n vertices has a Hamilton cycle or not. The vertices of the graph are labeled $1, 2, 3, \ldots, n$. The network has $\binom{n}{2}$ input gates $v_{i,j}$ ($1 \leq i < j \leq n$) and a single output gate u. The graph is specified by assigning "TRUE" to an input gate $v_{i,j}$ if and only if the vertices i and j are connected by an edge; the other input gates are assigned "FALSE". We want the output to be TRUE if and only if the graph has a Hamilton cycle, for all possible input graphs. Such a network can be designed, but the question is whether its size can be bounded by some polynomial in n, say by n^{100}. To understand the complexity of computations using graph theory is a BIGGGG challenge!

4 Probability

Around 1960, Paul Erdős and Alfréd Rényi developed the theory of random graphs. In their model, we start with n vertices, which we fix. Then we begin to add edges, where the location of each new edge is chosen randomly from all pairs of vertices that are not yet adjacent. After a certain prescribed number m of edges, we stop.

Of course, if we repeat this construction, we will very likely end up with a different graph. If n and m are large, however, the graphs constructed in this way will be very similar, with a very small probability of getting an "outlier" (this is a manifestation of the Law of Large Numbers). A related phenomenon is that in watching these random graphs develop (as edges are added), we observe sudden changes in their structure. For example, if we look at the graph when it has $m = 0.49n$ edges, it will almost surely consist of many small connected components. If we look at it again when it has $0.51n$ edges, then it will contain a single giant component (containing about 4% of all vertices, independently of n), along with a few very small ones (the sizes of which are small compared to n).

Determining typical properties of random graphs is not easy, and Erdős and Rényi worked out many. Less than a decade later, I learned probability theory from the lectures of Rényi, and he gave me copies of their papers on random graphs. I have to admit that I was not interested in them for a while. They contained long, detailed computations, and who likes to read such things? Since then, the field has blossomed into one of the most active areas in graph theory and has become fundamental for modeling the Internet. Of course I could not avoid working with them, as we shall see below.

Probability enters graph theory in other ways. In fact, it is becoming a fundamental tool in many areas of mathematics. Often questions may have nothing to do with probability, although their solutions involve random choice. Their proofs can be so simple and elegant that I can describe one of them here. Let H be a hypergraph. Under what conditions is it 2-colorable? For an ordinary graph, this is a classical question that can be answered easily (one possible answer is that the graph contain no odd cycles), but for a general hypergraph, this is a very hard question.

Erdős and Hajnal proved the following theorem around 1970: *If every edge of the hypergraph H has r vertices, and H has less than 2^{r-1} edges, then H is 2-colorable.* This is an example of a hypergraph question that "stands alone": the claim is trivial for graphs (you need 3 edges to create a non-bipartite graph), but very interesting for general hypergraphs.

Trying to prove this via usual methods (e.g. induction) does not work. But here is a proof using probability theory. Let us color the vertices at random: for each vertex, we flip a coin and color it red or blue depending on the outcome of the coin flip. There are possible "bad events": the event that a given edge has only one color is bad. If we are lucky, and none of these occur, then we get a good coloring. But can all bad events be avoided simultaneously? What is the probability of this?

Let us start with an easier task: what is the probability that a specific bad event is avoided? There are 2^r ways to color the vertices of an edge (the colors of the other vertices don't matter), and two of these are bad. So the probability that this edge is all red or all blue is $2/2^r = 2^{1-r}$.

Now the probability that one of these bad events happens is bounded by the sum of their probabilities. This is less than $2^{r-1} \times 2^{1-r} = 1$. Thus, good colorings must exist. It is difficult to construct a single good coloring, but coloring the vertices randomly works!

This method, called the *probabilistic method*, has become very powerful and important. I got involved in improving it. For example, I proved (working with Erdős) that in the above problem, we don't have to limit the number of edges; it suffices to assume that no edge meets more than 2^{r-3} other edges.

5 Algebra, Topology, and Graph Theory

Probability is not the only field of mathematics that has profound applications in graph theory. There are beautiful applications of very classical mathematics, such as algebra. I always found this fascinating, and tried to find connections myself. When I was a student, it seemed that mathematics was on a path towards fragmentation: different branches, and in particular relatively new branches like probability or graph theory, looked to be separating both from each other and from classical branches. I am happy to report, however, that this tendency seems to have turned around, and that the case for the unity of mathematics is on much firmer grounds.

This may sound like empty speculation, but I can illustrate such connections by citing a recent IMO problem, namely Problem 6 of 2007, and in particular its solution.

6 Networks, or Very Large Graphs

Many areas of mathematics, computer science, biology, physics, and the social sciences are concerned with properties of very large graphs (often called networks).

The Internet is an obvious example. There is in fact more than one network that we can define based on the Internet. One is a "physical" network, i.e. the graph whose vertices are electronic devices (computers, telephones, routers, hubs, etc.), and whose edges are the connections between them (wired or wireless). There is also a "logical" network, often called the World Wide Web, whose vertices are the documents available on the Internet, and whose (directed) edges are the hyperlinks that point from one to the other.

Social networks are of course formed by people, and they may be based on various definitions of connectivity. However the best known and best studied social networks are Internet-based (like Facebook). Some historians want to understand history based on a network of humans. The structure of this network determines, among other things, how fast news, disease, religion, and knowledge spread through society, and it has an enormous impact on the course of history.

There are many other networks related to humans. The brain is a great example of a huge network whose workings are not yet fully understood. It is too large for its structure (i.e. all of its neurons and their connections) to be encoded in our DNA. Why is it that it still functions and is able, for example, to solve math problems?

Biology is full of systems whose basic structure is a network. Consider the interactions between the plants and animals living in a forest (who eats whom?), or the interactions between proteins in our bodies. Networks are about to become part of a basic language for describing the systems and structures in many parts of nature — just as continuous functions and the operations of differentiation and integration are part of a basic language for describing mechanics and electromagnetism.

From the point of view of a mathematician, this should imply that powerful tools must be created to help biologists, historians, and sociologist describe the systems in which they (and all of us) are interested. This will not be an easy task, since these systems are very diverse. Modeling traffic, information distribution, and the electrical networks discussed above is only the tip of the iceberg.

Let me conclude with a few words about a topic that I have studied recently which is motivated by problems concerning very large graphs. The main idea is to assume that these graphs "tend to infinity" and to study the "limit objects". We often use the finite to approximate the infinite; obtaining numerical solutions to physical equations (say, for the purpose of predicting the weather) usually requires restricting space and time to a finite number of points, and then computing (more or less step-by-step) how temperature, pressure, etc. develop at these points.

The idea that the infinite may be a good approximation of the finite is more subtle. Continuous structures are often cleaner, more symmetric, and richer than their discrete counterparts.

To illustrate this idea with a physical example, consider a large piece of metal. This piece of metal is a crystal that is really a large graph consisting of atoms and bonds between them (arranged in a periodic and therefore rather boring way). But for an engineer who uses this metal to build a bridge, it is more useful to consider it as a continuum with a few important parameters (e.g. density, elasticity, and temperature) that are functions on this continuum. Our engineer can then use differential equations to compute the stability of the bridge. Can we regard a very large graph as some kind of a continuum?

In some cases this is possible, and Figure 2 illustrates the idea. We start with a random graph that is just a little more complicated than the random graphs introduced by Erdős and Rényi. It is constructed randomly according to the following rule: at each step, either a new vertex or a new edge is created. If the total number of vertices is n, then the probability that a new vertices is created is $1/n$, and the probability that a new edge is created is $(n-1)/n$. A new edge connects a randomly chosen pair of vertices.

The grid on the left represents the graph after 100 steps in the following way: the pixel at the intersection of the i-th row and the j-th column is black if there is an edge connecting the ith and j-th vertices, and white if there is no such edge. Thus the area in the upper left is darker, because a pixel there represents a pair of vertices that have been around for a longer time and hence have a greater chance of being connected.

2 A randomly grown uniform attachment graph with 100 vertices, and the continuous function $1 - \max(x,y)$ approximating it.

Although this graph is random, the pixel picture on the left is, from a distance, quite similar to the continuous function $1 - \max(x,y)$, which is depicted on the right. If instead of a hundred steps we took 1000, the similarity would be even more striking. One can prove that the rather simple function on the right encodes all of the information one needs to know about the graph on the left, except for random fluctuations, which become smaller and smaller as the number of vertices grows.

Large graphs, with thousands of vertices, and huge graphs, with billions, represent a new kind of challenges for the graph theorist. It hints at the beauty of mathematics that to meet some of these challenges we have to discover and use more and more connections with other, more classical parts of mathematics.

7.4 Stanislav Smirnov

Geneva, Switzerland

Stanislav Smirnov (* 1970) grew up in Sankt Petersburg, in the Soviet Union at the time. He participated twice at IMOs, in 1986 (Poland) and 1987 (Cuba), and both times earned a perfect score of 42 points. He studied at Sankt Petersburg State University and obtained his PhD in 1996 at the California Institute of Technology; his supervisor was Nikolai Makarov.

He held positions at Yale, Bonn, and Princeton, before moving the the Royal Institute of Technology in Stockholm in 1998. Since 2003, he has been Professor of Mathematics at the University of Geneva.

In 2001, Stanislav Smirnov won the Salem Prize (together with Oded Schramm) as well as a Clay Research Award; in 2004, he was awarded the European Mathematical Society Prize.

The research areas of Stanislav Smirnov are in complex and geometric analysis, dynamical systems (in particular complex dynamics), and in probability theory. His initial work, including his thesis research, was on complex dynamics (dynamical properties of Julia sets).

Stanislav Smirnov is most famous for his work on percolation theory. This theory is coming from mathematical physics, but is used in a number of applied and engineering areas (including oil drilling and coffee brewing); it has been an area of active mathematical research, fundamental and applied. Stanislav Smirnov is among those who contributed most towards a successful mathematical theory.

The basic setting in percolation theory is a triangular grid in the plane. Each edge in this grid can have two states, say "open" and "closed", and all edges have equal and independent probability, say p, for being open (and otherwise closed). One is interested in paths along open edges, and clusters consisting of open edges. (This could model for instance conductivity in porous media, where "open" edges describe permeable spots in some substance, and "closed" edges are impermeable.) One question concerns the average size of clusters of red vertices, depending on the probability p. There is a critical probability p_{crit} at which the average cluster size jumps from finite to infinite.

One fundamental question that plays a significant role in mathematical physics is that of *conformal invariance*. Given a finite domain U in the plane that is covered by a triangular grid. Then the domain is deformed by a smooth map of the plane that preserves all angles (such maps are called *conformal*). The image domain is covered again by a triangular grid. Are the percolation properties in the original and the image domains, with their two respective grids, the same? This is, loosely speaking, the statement of conformal invariance of percolation, and this had been a fundamental conjecture in the field. Smirnov proved this result in 2001, along with "Cardy's formula" that determines the probability that two boundary segments of a bounded domain can be connected by a path along open edges. It remains an open question whether there is an analogous result for square grids, rather than triangular grids.

How do Research Problems Compare with IMO Problems?

Stanislav Smirnov

1 Do Mathematicians Solve Problems?

When asked what research in mathematics is like, mathematicians often answer: *We prove theorems.* This best describes the quintessential part of mathematical work and also how it differs from research in, say, biology or linguistics. And though in school one often gets the impression that all theorems were proved ages ago by Euclid and Pythagoras, there are still many important unsolved problems.

Indeed, research mathematicians do solve problems. There are other important parts of research, from learning new subjects and looking for connections between different areas to introducing new structures and concepts and asking new questions. Some even say that posing a problem is more important than solving it. In any case, without problems there would be no mathematics, and solving them is an important part of our job. As Paul Halmos, who has written several books about research problems, once said: *Problems are the heart of mathematics.*

Students often ask: *How does doing research compare to the IMO experience?* There are many similarities, and problem solving skills certainly help in research, so many IMO competitors go on to become mathematicians. There are also, however, some differences. So how do IMO problems compare to research problems?

Substantial differences in solutions are often mentioned. Typically, an IMO problem will have a nice solution that requires the use of a limited number of methods (and hopefully at least one participant can find it within the given four and a half hours). Problems one encounters as a mathematician often require methods from very different mathematical areas, so that ingenuity alone would not suffice to solve them. Moreover for many problems that are easy to formulate only long and technical solutions have been found; it may even be the case that no nice solution exists. And when starting to work on a research problem, you

Stanislav Smirnov
Section de Mathématiques, Université de Genève, 2–4 rue du Lièvre, CP 64,
1211 Genève 4, Suisse. e-mail: Stanislav.Smirnov@unige.ch

cannot be sure that there is a solution at all. So you do not need to be as quick as when at IMO competitions, but you need to have much more determination — you rarely prove a theorem in four hours, and sometimes it takes years to advance on an important question. On the positive side, mathematics is now more of a collective effort, and collaborating with others is a very rewarding experience.

Not only the solutions, but also the problems themselves are somewhat different. Three IMO problems easily fit on one sheet of paper, but it takes much more to describe most of the open questions in mathematical research. Fortunately, there are exceptions that have equally short formulations, and mathematicians very much enjoy tackling these — they often play a catalytic role, attracting our attention to a particular area. Motivation for posing problems also differs. Whereas many research problems (like most of IMO problems) are motivated by the inner beauty of mathematics, a significant number originate in physical or practical applications, and then the questions asked change somewhat.

So, are research problems and IMO problems really different? I would say that they have more in common, and that mathematicians enjoy beautiful problems, elegant solutions, and the process of working on a problem just as much as IMO contestants do.

To highlight both the similarities and the differences, I describe below a few problems that I have encountered and that would do well as both IMO and research questions (but would come in a slightly different light). While coming from different areas of mathematics, all of these are concerned with numbers (or colors) placed on a graph.

2 The Pentagon Game

The Pentagon Game is one of the most memorable problems I solved at olympiads. It was proposed by Elias Wegert of Germany, who also took part in the 50-th IMO as a coordinator:

> ## 27th International Mathematical Olympiad
> Warsaw, Poland
> Day I
> July 9, 1986
>
> **Problem 3.** *To each vertex of a regular pentagon an integer is assigned in such a way that the sum of all five numbers is positive. If three consecutive vertices are assigned the numbers x, y, z respectively and $y < 0$ then the following operation is allowed: the numbers x, y, z are replaced by $x + y$, $-y$, $z + y$ respectively. Such an operation is performed repeatedly as long as at least one of the five numbers is negative. Determine whether this procedure necessarily comes to an end after a finite number of steps.*

I was among the students, and it was a very nice problem to tackle, perhaps the hardest at that IMO. It is almost immediately clear that one should find some positive integer function of a configuration that decreases with each operation. Indeed, two such semi-invariants were found by participants, and since we cannot decrease a positive integer infinitely many times, the procedure will necessarily come to an end.

This is a classical *combinatorics* problem, and if you are into olympiads, you certainly have seen a few very similar ones. What is interesting is that its life was more like that of a research problem. It was originally motivated by a question that arose in research dealing with partial reflections of polygons. So even the motivating area, *geometry*, was very different.

The combinatorial structure of this game is interesting in itself, and studying it on graphs that are different from a pentagon could have led to a few IMO problems and perhaps a research paper. But connections with algebraic questions have surfaced, which made it much more interesting for mathematical research. I was very pleasantly surprised to hear a talk that originated from the Pentagon Game at a research seminar some twenty years after that IMO. The talk was by Qëndrim Gashi, who used a version of the game due to Shahar Mozes to prove the Kottwitz-Rapoport conjecture in *algebra*. So far, versions of the Pentagon Game have led to more than a dozen research papers — not bad for an IMO problem!

These kinds of unexpected links between different areas, and between simple and complicated subjects, are one of the best things about doing mathematical research. Unfortunately, they often pass unnoticed in IMO competitions.

3 The Game of Life

There are many similar games with numbers, and they may yield much wider connections, often stretching beyond mathematics.

The most famous is perhaps John Conway's *Game of Life*. This is an example of a very rich class of games called cellular automata, first introduced by John von Neumann and Stanisław Ulam. In such games the graph is taken to be a regular grid, only a finite set of numbers (or states) is used, and an operation consists in *simultaneously* changing all numbers according to some rule depending on their neighbors.

The Game of Life is played on a square grid, with squares that can have two states: 0 (black) and 1 (white). The operation simultaneously changes the states of all squares by a simple rule depending on the state of their eight neighbors (i.e. squares with which they share an edge or a vertex):

- a black cell with 2 or 3 black neighbors becomes black,
- a white cell with 3 black neighbors becomes black,
- all other cells become white.

The rule is very simple, but it leads to rather complicated phenomena. Besides configurations that stay fixed (e.g. a 2×2 square of live cells) and those that oscillate periodically (e.g. a 1×3 rectangle of live cells), there are configurations that exhibit nontrivial behavior. For example, the "glider" pattern moves one step southeast every four operations, whereas "Bill Gosper's gun" shoots out a new glider every thirty operations. Patterns like this allow us to use the Game of Life to even model a computer, though the needed configurations would be rather large and complicated. Also, chaotic configurations quite often transform into complex patterns with some structure, which makes the game interesting to scientists in other disciplines, from philosophy to economics.

The Game of Life was popularized by Martin Gardner well beyond the mathematics community, and one can now easily find information about it, including interactive models, on the internet. Moreover many questions about this game could well double as IMO and research problems, and there are many other interesting cellular automata.

4 The Sandpile Model

It seems natural that if one wants to model more phenomena, some randomness has to be added, so that the evolution is not uniquely determined by the initial configuration. Indeed, it has been known for a long time that by introducing randomness into simple games (roughly speaking, we take two or more rules, and at each vertex toss a coin to decide which one to apply) one can accurately model many phenomena exhibiting phase transitions — from ferromagnetic materials to the spread of epidemics. It came as quite a surprise, however, that similar phenomena could be observed even in usual, non-random games.

One famous such game, the *Sandpile Model*, was introduced by three physicists, Per Bak, Chao Tang, and Kurt Wiesenfeld, in 1987. The game is played on an infinite square grid by writing positive integers on finitely many cells, and zeroes elsewhere. These integers are thought of as heights of a pile of sand. (One can also play on a finite region, but then one cell is designated a "pit": all grains falling there disappear.)

In the original model, all cells were changed simultaneously. Below we give a modified version, due to Deepak Dhar, where the same rule is applied but only to one cell at a time, much like in the Pentagon Game. The operation is slightly different, though: whereas in the Pentagon Game we subtract $2y$ from a vertex with value y and redistribute this amount evenly among neighbors, here we subtract 4. To be precise, the operation in the Sandpile Model is the following: if some cell with h grains is too tall (i.e. $h \geq 4$), it will topple over, giving one grain of sand to each of its four neighboring cells (i.e. those cells with which it shares an edge), which have, say, h_1, h_2, h_3, and h_4 grains. Thus the operation is described by

$$h \rightarrow h-4\,,$$
$$h_j \rightarrow h_j+1\,.$$

Like in the Pentagon Game, the operation is performed repeatedly as long as we can find a cell with $h \geq 4$. Eventually, we stop, reaching a stable configuration with all piles satisfying $h \leq 3$. The sequence of operations that leads to a stable configuration is called an *avalanche*.

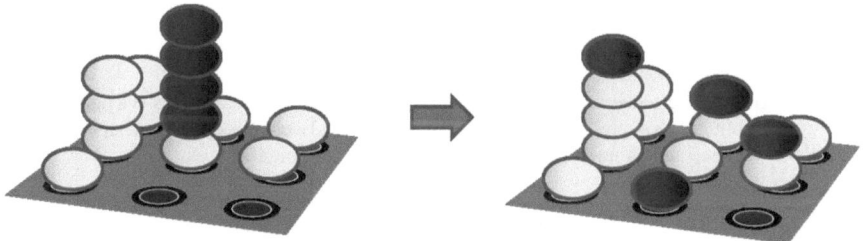

1 A pile of five grains is toppled, with one grain going to each of its four neighbors. Note that we created a new pile of four grains, ready to be toppled. (The figure shows a 3×3 square within the infinite square grid.)

In order to work with the sandpile model, one first has to solve a problem very similar to one from the 1986 IMO:

Show that an avalanche ends after a finite number of operations.

Often, more than one pile has big height, so that we must choose the pile that we wish to topple. But it appears (unlike with the Pentagon Game) that

at the end of an avalanche, one obtains the same configuration independently of the order of operations.

2 50000 grains were added at the central square, resulting in an avalanche. This is a pile obtained afterwards, with colors representing the heights (0, 1, 2 or 3) of the cells. The shape is almost circular. If we keep adding more grains, will it look more like a circle?

Can you prove these two statements? In addition to being important lemmas in research papers, they would make for nice IMO problems.

According to physicists (and we have high respect for our colleagues — interactions between mathematics and physics have enriched both fields), the really interesting problems only start here. Once the avalanche comes to an end, we can add one more grain of sand at some fixed center cell (or at a random place). This leads to a new avalanche. Then we add a new grain, and so forth.

When the original sandpile paper appeared, physicists were struggling to explain two recurring phenomena in nature: the "$1/f$ noise" and the appearance of spatial fractal structures. Both phenomena are often encountered in everyday life: $1/f$-noise (so called because its power is inversely proportional to its frequency) appears in areas as different as hissing sounds of a home stereo system, human heartbeats, or stock market fluctuations. And seemingly chaotic yet self-similar fractal structures (so called because they behave as if their dimension was fractional) can be seen in the shapes of clouds, systems of blood vessels, or mountain ranges. Based on physical observations, one could ask questions like: given a pile totaling N grains, what would be the average diameter (size), length (number of grains toppled), or shape of an avalanche?

3 An avalanche triggered by adding one grain (not at the center) to the pile of 50 000 grains from Figure 4. What would be the average size of such an avalanche?

Computer experiments have exposed both phenomena in the Sandpile Model: adding grains to a stable configuration triggers an avalanche of a fractal shape and a size distributed not unlike $1/f$ noise. Moreover it is usually the case that adding one grain of sand either does not change much or causes almost the entire pile to collapse in an avalanche. Such behavior is characteristic of physical systems at "critical points", like a liquid around freezing temperature, when a small change (slightly decreasing the temperature, or sometimes dropping a small crystal into it) can cause it to freeze. However, the Sandpile Model is attracted to the critical point, whereas most physical systems are difficult to keep at criticality. And the sandpile model, though very simple to formulate, turned out to be the first mathematical example of what physicists now call "self-organized criticality".

Despite extensive computer simulations providing convincing evidence, as well as a vast literature, most of these questions remain open after 20 years, but mathematicians work for much longer than four and a half hours! Also, it is not clear that such questions will have a nice (and provable) answer, and their original motivation comes from outside of mathematics, so that they would likely not be asked as IMO problems.

Though the original motivation for the Sandpile Model came from physics, mathematicians have since asked a number of mathematical questions about it, motivated by the simplicity and beauty of the model. Some of these questions are of geometric nature and would do fine as IMO problems if only we had nice solutions. For example, we can keep adding particles at the origin, and the pile will grow in size. Will it look like a circle, as Figure 4 suggests? Apparently not — it seems that it will start developing sides after a while. So what will be its shape? How can we describe the intricate patterns we see? Despite much work, we still do not know.

5 The Self-Avoiding Walk

This article started with a problem that I solved at the 1986 IMO. It therefore seems appropriate to end with one of the problems that I am trying to solve now. Lately, I have spent a lot of time working on a large class of questions about systems undergoing phase transitions, and they too can be formulated as games played on a grid. The problem below is perhaps the simplest to formulate, and one does not even need to define a game, just counting configurations on a grid is enough.

In the 1940s, Paul Flory, a Nobel Prize winning chemist, asked how a polymer is positioned in space. He proposed modeling polymer chains by broken lines drawn on a grid without self-intersections (since a molecule obviously won't intersect itself). Equivalently, imagine a person walking on a grid in such a way that he does not visit the same vertex twice. We call this a *self-avoiding walk*. Each n-step trajectory would then model a possible position of a length n chain.

The fundamental question is what generic chains would look like, but before answering this one has to ask the following question:

> How many length n self-avoiding walks starting from the origin can one draw on a grid?

Denote this number by $C(n)$; walks that are rotations of each other are counted separately. It depends on the grid chosen, and in general we do not expect to have a nice formula for it (though it may exist despite expectations — sometimes miracles do happen). One thus asks how fast this number grows in terms of n. An IMO-type problem might be the following:

> Show that there is a constant μ such that the number of self-avoiding walks satisfies $C(n) \approx \mu^n$ as n tends to ∞.

The sign $\approx$ above means that however small we take ε, for large enough n we have $(\mu - \varepsilon)^n < C(n) < (\mu + \varepsilon)^n$. The problem above is not difficult and follows

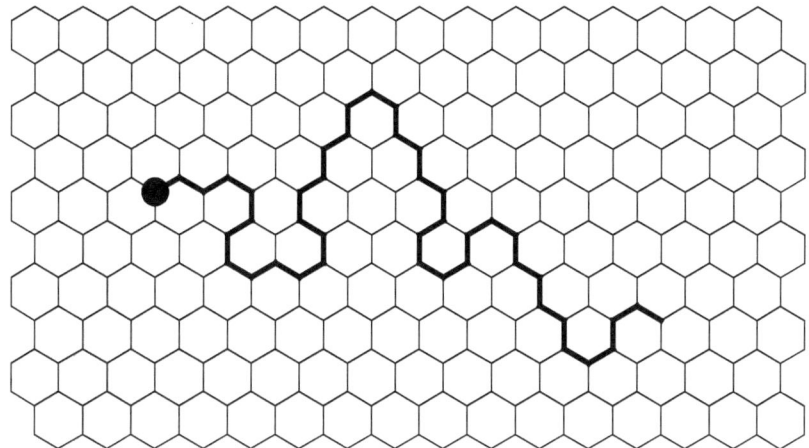

4 A self-avoiding walk on the hexagonal lattice. Starting from the origin, we make steps along edges, visiting each vertex at most once. How many such walks of length n are there?

from the observation that a self-avoiding walk of length $n + m$ can be cut into two self-avoiding walks of lengths n and m, and hence $C(n + m) \leq C(n) \cdot C(m)$.

On the other hand, if we glue together two self-avoiding walks of lengths n and m, the resulting walk of length $n + m$ need not be self-avoiding, so that in general there is no equality. This makes the determination of $C(n)$ difficult.

The number μ (which depends on the grid) is called the *connective constant* and has several important applications, so its determination is indeed important.

Quite some time passed before a guess was even made about the actual values. In 1982, the physicist Bernard Nienhuis found a heuristic argument for the value of μ on the two-dimensional hexagonal lattice, like the one in Figure 4. His argument strongly suggested that in this case

$$C(n) \approx \left(\sqrt{2 + \sqrt{2}} \right)^n ,$$

and moreover, if one wants higher precision, that for every $\varepsilon > 0$

$$\left(\sqrt{2 + \sqrt{2}} \right)^n n^{\frac{11}{32} - \varepsilon} < C(n) < \left(\sqrt{2 + \sqrt{2}} \right)^n n^{\frac{11}{32} + \varepsilon} ,$$

when n is large. His arguments are quite beautiful and inspiring, but they are not mathematically rigorous (and would not have yielded a full score if submitted at a mathematical olympiad).

More than 20 years had to pass before a mathematical solution confirming this prediction was found. Just two months after giving a talk on this subject at the IMO anniversary in Bremen, together with Hugo Duminil-Copin we proved that on a hexagonal lattice indeed $\mu = \sqrt{2 + \sqrt{2}}$. Surprisingly, the proof is elementary, and so short that it could be written within the IMO time limit! We carefully count the self-avoiding walks, looking not only at their lengths but also at their windings (i.e. the number of turns they make). The value of μ arises in relation to turns, as $2\cos(\pi/8)$.

So is it surprising that the proof had to wait for so long, and could this problem be given as an IMO problem? The answer to both questions is no: although our way of counting is elementary, inventing it the first time around is far from easy, and requires knowing, in addition to mathematics, a great deal of physics. And what about $11/32$? We are probably closer to it than ever before. The mathematicians Greg Lawler, Oded Schramm and Wendelin Werner have explained where it would come from, and together our results could eventually lead to the proof. But it is still some months or years away, and probably won't be elementary — surprisingly, the number $11/32$ arises in a much more complicated way than $2\cos(\pi/8)$!

6 Conclusion

Problem solving experience gained at the IMO will be useful to you regardless of what you decide to do in life. But it will be especially useful if you decide to become a mathematician, and although there are many other things involved in mathematical research besides problem solving, most of them are exciting as well. Mathematics is currently an exciting field, with many beautiful problems and many surprising connections between different branches and to other disciplines. It has become a truly collaborative effort, and is as international as the IMOs — the researchers mentioned in this short article alone come from almost a dozen different countries. I hope that many IMO participants will go on to become mathematicians, and that we will meet again.

7.5 Terence Tao

Los Angeles, USA

Terence Tao's vitae is peppered with superlatives, usually in combination with "youngest". Reportedly having learned arithmetic at the age of two by watching Sesame Street, he won a bronze medal at the 1986 IMO at the age of ten — the youngest IMO participant to date — and silver in 1987 and gold in 1988 — another "youngest".

Today, at the age of 34, he is the youngest senior mathematician at UCLA (University of California at Los Angeles), having been full professor for almost ten years.

He has received many of the most prestigious awards in mathematics and science, including the Salem Price, a MacArthur Fellowship, and the Fields Medal.

The strength of Tao's work comes from his superb ability to draw connections between different branches of mathematics. Though his research career started out in harmonic analysis and partial differential equations, he has brought in ideas from probability and ergodic theory and made substantial progress in areas of combinatorics, compressed sensing, representation theory, and number theory — most notably what is now known as the Green–Tao theorem proving existence of arbitrarily long arithmetic progressions of prime numbers.

Tao embodies the modern-day version of the traveling mathematician. Before the advent of the world wide web, Paul Erdős, arguably the icon of 20th century mathematics, traveled constantly to maintain his many simultaneous collaborations. Tao's solution is his research weblog which, besides Tao's own prolific and lucid posts, attracts contributions from known and not-so-known mathematicians which transpire seriousness as well as the fun of doing mathematics. In addition to his strong commitment to traditional collaborative research — in his own words, "those papers that I consider among my best work, they are virtually all joint" — he is actively participating in massively collaborative research initiatives. With a combination of transparency, openness, and quality, Tao's impact goes beyond his major contributions to our mathematical knowledge, influencing the very process of how mathematics is done in the 21st century.

Structure and Randomness in the Prime Numbers

Terence Tao

Abstract. A quick tour through some topics in analytic prime number theory.

1 Introduction

The prime numbers $2, 3, 5, 7, \ldots$ are one of the oldest topics studied in mathematics. We now have a lot of intuition as to how the primes *should* behave, and a great deal of confidence in our conjectures about the primes... but we still have a great deal of difficulty in *proving* many of these conjectures! Ultimately, this is because the primes are believed to behave *pseudorandomly* in many ways, and not to follow any simple pattern. We have many ways of establishing that a pattern exists... but how does one demonstrate the *absence* of a pattern?

In this article I will try to convince you why the primes are believed to behave pseudorandomly, and how one could try to make this intuition rigorous. This is only a small sample of what is going on in the subject; I am omitting many major topics, such as sieve theory or exponential sums, and am glossing over many important technical details.

2 Finding Primes

It is a paradoxical fact that the primes are simultaneously very numerous, and hard to find. On the one hand, we have the following ancient theorem [2]:

Theorem 7.1 (Euclid's Theorem). *There are infinitely many primes.*

Terence Tao
Department of Mathematics, UCLA, Los Angeles CA 90095-1555.
e-mail: tao@math.ucla.edu

In particular, given any k, there exists a prime with at least k digits. But there is no known *quick* and *deterministic* way to locate such a prime! (Here, "quick" means "computable in a time which is polynomial in k".) In particular, there is no known (deterministic) formula that can quickly generate large numbers that are guaranteed to be prime. Currently, the largest known prime is $2^{43,112,609} - 1$, about 13 million digits long [3].

On the other hand, one can find primes quickly by *probabilistic* methods. Indeed, any k-digit number can be tested for primality quickly, either by probabilistic methods [9, 11] or by deterministic methods [1]. These methods are based on variants of Fermat's little theorem, which asserts that $a^n \equiv a \mod n$ whenever n is prime. (Note that if n is a k-digit number, $a^n \mod n$ can be computed quickly, by first repeatedly squaring a to compute $a^{2^j} \mod n$ for various values of j, and then expanding n in binary and multiplying the indicated residues $a^{2^j} \mod n$ together.)

Also, we have the following fundamental theorem [8, 13]:

Theorem 7.2 (Prime Number Theorem). *The number of primes less than a given integer n is $(1 + o(1))\frac{n}{\log n}$, where $o(1)$ tends to zero as $n \to \infty$.*

(We use log to denote the natural logarithm.) In particular, the probability of a randomly selected k-digit number being prime is about $\frac{1}{k \log 10}$. So one can quickly find a k-digit prime with high probability by randomly selecting k-digit numbers and testing each of them for primality.

Is randomness really necessary?   To summarize: We do not know a quick way to find primes *deterministically*. However, we have quick ways to find primes *randomly*.

On the other hand, there are major conjectures in complexity theory, such as $P = BPP$, which assert (roughly speaking) that any problem that can be solved quickly by probabilistic methods can also be solved quickly by deterministic methods.

These conjectures are closely related to the more famous conjecture $P \neq NP$, which is a USD \$ 1 million Clay Millennium prize problem.

Many other important probabilistic algorithms have been *derandomised* into deterministic ones, but this has not been done for the problem of finding primes. (A massively collaborative research project is currently underway to attempt this [10].)

3 Counting Primes

We've seen that it's hard to get a hold of any single large prime. But it is easier to study the set of primes *collectively* rather than one at a time.

An analogy: it is difficult to locate and count all the grains of sand in a box, but one can get an estimate on this count by *weighing* the box, subtracting the weight of

---

Strictly speaking, the $P = BPP$ conjecture only applies to *decision problems* — problems with a yes/no answer —, rather than *search problems* such as the task of finding a prime, but there are variants of $P = BPP$, such as $P = $ promise-*BPP*, which would be applicable here.

the empty box, and dividing by the average weight of a grain of sand. The point is that there is an easily measured statistic (the weight of the box with the sand) which reflects the *collective* behaviour of the sand.

For instance, from the *fundamental theorem of arithmetic* one can establish *Euler's product formula*

$$\sum_{n=1}^{\infty} \frac{1}{n^s} = \prod_{p \text{ prime}} \left(1 + \frac{1}{p^s} + \frac{1}{p^{2s}} + \frac{1}{p^{3s}} + \cdots \right) = \prod_{p \text{ prime}} \left(1 - \frac{1}{p^s} \right)^{-1} \qquad (1)$$

for any $s > 1$ (and also for other complex values of s, if one defines one's terms carefully enough).

The formula (1) links the collective behaviour of the primes to the behaviour of the *Riemann zeta function*

$$\zeta(s) := \sum_{n=1}^{\infty} \frac{1}{n^s} \,,$$

thus

$$\prod_{p \text{ prime}} \left(1 - \frac{1}{p^s} \right) = \frac{1}{\zeta(s)} \,. \qquad (2)$$

One can then deduce information about the primes from information about the zeta function (and in particular, its zeroes).

For instance, from the divergence of the harmonic series $\sum_{n=1}^{\infty} \frac{1}{n} = +\infty$ we see that $\frac{1}{\zeta(s)}$ goes to zero as s approaches 1 (from the right, at least). From this and (2) we already recover Euclid's theorem (Theorem 7.1), and in fact obtain the stronger result of Euler that the sum $\sum_p \frac{1}{p}$ of reciprocals of primes diverges also.

In a similar spirit, one can use the techniques of complex analysis, combined with the (non-trivial) fact that $\zeta(s)$ is never zero for $s \in \mathbb{C}$ when $\mathrm{Re}(s) \geq 1$, to establish the prime number theorem (Theorem 7.2); indeed, this is how the theorem was originally proved [8, 13] (and one can conversely use the prime number theorem to deduce the fact about the zeroes of ζ).

The famous *Riemann hypothesis* asserts that $\zeta(s)$ is never zero when $\mathrm{Re}(s) > 1/2$. It implies a much stronger version of the prime number theorem, namely that the number of primes less than an integer $n > 1$ is given by the more precise formula $\int_0^n \frac{dx}{\log x} + O(n^{1/2} \log n)$, where $O(n^{1/2} \log n)$ is a quantity which is bounded in magnitude by $C n^{1/2} \log n$ for some absolute constant C (for instance, one can take $C = \frac{1}{8\pi}$ once n is at least 2657 [12]). The hypothesis has many other consequences in number theory; it is another of the USD \$ 1 million Clay Millennium prize problems. More generally, much of what we know about the primes has come from an extensive study of the properties of the Riemann zeta function and its relatives, al-

---

A technical point: the sum $\sum_{n=1}^{\infty} \frac{1}{n^s}$ does not converge in the classical sense when $\mathrm{Re}(s) \leq 1$, so one has to interpret this sum in a fancier way, or else use a different definition of $\zeta(s)$ in this case; but I will not discuss these subtleties here.

though there are also some questions about primes that remain out of reach even assuming strong conjectures such as the Riemann hypothesis.

4 Modeling Primes

A fruitful way to think about the set of primes is as a *pseudorandom set* — a set of numbers which is not actually random, but behaves like one.

For instance, the prime number theorem asserts, roughly speaking, that a randomly chosen large integer n has a probability of about $1/\log n$ of being prime. One can then *model* the set of primes by replacing them with a random set of integers, in which each integer $n > 1$ is selected with an independent probability of $1/\log n$; this is *Cramér's random model*.

This model is too crude, because it misses some obvious structure in the primes, such as the fact that most primes are odd. But one can improve the model to address this, by picking a model where odd integers n are selected with an independent probability of $2/\log n$ and even integers are selected with probability 0.

One can also take into account other obvious structure in the primes, such as the fact that most primes are not divisible by 3, not divisible by 5, etc. This leads to fancier random models which we believe to accurately predict the asymptotic behaviour of primes.

For example, suppose we want to predict the number of twin primes n, $n+2$, where $n \leq N$ for a given threshold N. Using the Cramér random model, we expect, for any given n, that $n, n+2$ will simultaneously be prime with probability $\frac{1}{\log n \log(n+2)}$, so we expect the number of twin primes to be about

$$\sum_{n=1}^{N} \frac{1}{\log n \log(n+2)} \approx \frac{N}{\log^2 N}.$$

This prediction is inaccurate; for instance, the same argument would also predict plenty of pairs of *consecutive* primes $n, n+1$, which is absurd. But if one uses the refined model where odd integers n are prime with an independent probability of $2/\log n$ and even integers are prime with probability 0, one gets the slightly different prediction

$$\sum_{\substack{1 \leq n \leq N \\ n \text{ odd}}} \frac{2}{\log n} \times \frac{2}{\log(n+2)} \approx 2\frac{N}{\log^2 N}.$$

More generally, if one assumes that all numbers n divisible by some prime less than a small threshold w are prime with probability zero, and are prime with a probability of $\prod_{p<w}(1-\frac{1}{p})^{-1} \times \frac{1}{\log n}$ otherwise, one is eventually led to the prediction

---

We use the symbol $\approx$ in the sense that the quotient of the two quantities tends to 1 as $N \to \infty$.

$$2 \left(\prod_{\substack{p<w \\ p \text{ odd}}} \frac{p-2}{p} \left(1-\frac{1}{p}\right)^{-2} \right) \frac{N}{\log^2 N} = 2 \left(\prod_{\substack{p<w \\ p \text{ odd}}} \left(1-\frac{1}{(p-1)^2}\right) \right) \frac{N}{\log^2 N}$$

(for p an odd prime, among p consecutive integers, only $p-2$ have a chance to be the smaller number in a pair of twin primes). Sending $w \to \infty$, one is led to the asymptotic prediction

$$\Pi_2 \frac{N}{\log^2 N}$$

for the number of twin primes less than N, where Π_2 is the *twin prime constant*

$$\Pi_2 := 2 \prod_{p \text{ odd prime}} \left(1-\frac{1}{(p-1)^2}\right) \approx 1.32032\ldots.$$

For $N = 10^{10}$, this prediction is accurate to four decimal places, and is believed to be asymptotically correct. (This is part of a more general conjecture, known as the *Hardy-Littlewood prime tuples conjecture*.)

Similar arguments based on random models give convincing heuristic support for many other conjectures in number theory, and are backed up by extensive numerical calculations.

5 Finding Patterns in Primes

Of course, the primes are a deterministic set of integers, not a random one, so the predictions given by random models are not rigorous. But can they be made so?

There has been some progress in doing this. One approach is to try to classify all the possible ways in which a set could *fail* to be pseudorandom (i.e. it does something noticeably different from what a random set would do), and then show that the primes do not behave in any of these ways.

For instance, consider the *odd Goldbach conjecture*: every odd integer larger than five is the sum of three primes. If, for instance, all large primes happened to have their last digit equal to one, then Goldbach's conjecture could well fail for some large odd integers whose last digit was different from three. Thus we see that the conjecture could fail if there was a sufficiently strange "conspiracy" among the primes.

However, one can rule out this particular conspiracy by using the *prime number theorem in arithmetic progressions*, which tells us that (among other things) there are many primes whose last digit is different from 1. (The proof of this theorem is based on the proof of the classical prime number theorem.)

Moreover, by using the techniques of *Fourier analysis* (or more precisely, the *Hardy-Littlewood circle method*), we can show that *all* the conspiracies which could conceivably sink Goldbach's conjecture (for large integers, at least) are broadly of this type: an unexpected "bias" for the primes to prefer one remainder modulo 10 (or modulo another base, which need not be an integer), over another.

Vinogradov [14] eliminated each of these potential conspiracies, and established *Vinogradov's theorem*: every sufficiently large odd integer is the sum of three primes. This method has since been extended by many authors, to cover many other types of patterns; for instance, related techniques were used by Ben Green and myself [4] to establish that the primes contain arbitrarily long arithmetic progressions, and in subsequent work of Ben Green, myself, and Tamar Ziegler [5, 6, 7] to count a wide range of other additive patterns also. (Very roughly speaking, known techniques can count additive patterns that involve two independent parameters, such as arithmetic progressions $a, a+r, \ldots, a+(k-1)r$ of a fixed length k.)

Unfortunately, "one-parameter" patterns, such as twins $n, n+2$, remain stubbornly beyond current technology. There is still much to be done in the subject!

Recommended Reading

1. Manindra Agrawal, Neeraj Kayal, and Nitin Saxena, *PRIMES is in P*. Annals of Mathematics (2) **160** (2004), 781–793.
2. Euclid, *The Elements*, circa 300 BCE.
3. Great Internet Mersenne Prime Search, 2008; http://www.mersenne.org .
4. Ben Green and Terence Tao, *The primes contain arbitrarily long arithmetic progressions*. Annals of Mathematics **167** 2 (2008), 481–547.
5. Ben Green and Terence Tao, *Linear equations in primes*. Preprint, April 22, 2008, 84 pages; http://arxiv.org/abs/math/0606088 .
6. Ben Green and Terence Tao, *The Möbius function is asymptotically orthogonal to nilsequences*. Preprint, April 26, 2010, 22 pages; http://arxiv.org/abs/0807.1736 .
7. Ben Green, Terence Tao, and Tamar Ziegler, *The inverse conjecture for the Gowers norm*, preprint.
8. Jacques Hadamard, *Sur la distribution des zéros de la fonction $\zeta(s)$ et ses conséquences arithmétiques*. Bulletin de la Société Mathématique de France **24** (1896), 199–220.
9. Gary L. Miller, *Riemann's hypothesis and tests for primality*. Journal of Computer and System Sciences **13** 3 (1976), 300–317.
10. Polymath4 project: Deterministic way to find primes; http://michaelnielsen.org/polymath1/index.php?title=Finding_primes .
11. Michael O. Rabin, *Probabilistic algorithm for testing primality*. Journal of Number Theory **12** (1980), 128–138.
12. Lowell Schoenfeld, *Sharper bounds for the Chebyshev functions $\theta(x)$ and $\psi(x)$. II*. Mathematics of Computation **30** (1976), 337–360.
13. Charles-Jean de la Vallée Poussin, *Recherches analytiques de la théorie des nombres premiers*. Annales de la Société scientifique de Bruxelles **20** (1896), 183–256.
14. Ivan M. Vinogradov, *The method of trigonometrical sums in the theory of numbers* (Russian). Travaux de l'Institut Mathématique Stekloff **10** (1937).

7.6 Jean-Christophe Yoccoz

Paris, France

Jean-Christophe Yoccoz (* 1957) participated twice at IMOs, in 1973 (USSR) and 1974 (East Germany). From the USSR, he returned with a Silver medal; in East Germany he earned the first-ever Gold medal for France, tied on the first place.

Jean-Christophe Yoccoz studied Mathematics at the École Normale Supérieure in Paris, and obtained his PhD at the École Polytechnique in Palaiseau, south of Paris, under the direction of Michel Herman, in 1985.

In 1991, he obtained the Salem prize, and in 1994 he was awarded the Fields medal at the International Congress of Mathematicians in Zürich, Switzerland. He is professor of Mathematics at the Université Paris XI (Paris-Sud) in Orsay, as well as at the Collège de France in Paris. Jean-Christophe Yoccoz is a member of the French and Brazilian Academies of Science.

The research area of Jean-Christophe Yoccoz is dynamical systems, often in interplay with complex analysis, number theory, and other areas of mathematics. Probably the founding question of dynamical systems is this: *is the solar system stable?* (Will the planets continue to circle indefinitely around the sun, or will the mutual interaction forces eventually cause some of the planets to crash into the sun, or to be ejected from the solar system — under the idealized assumption that the sun remains the way it is.) One of the key questions of dynamical systems thus concerns its stability under small perturbations that act for a long time.

One of the most prominent results of Yoccoz is related to this question: he was the first to prove an explicit necessary and sufficient condition for a particular class of dynamical systems to be dynamically stable. His result concerns the simplest non-trivial class of dynamical systems, (complex) quadratic polynomials; the explicit condition involves detailed number-theoretic properties of rotation angles. Even though this result is on a quite specific class of "toy-model" maps, it has impact on dynamical systems in much greater generality, using "universality properties" (renormalization theory). This is also the topic that he will present at the IMO anniversary ceremony.

Another deep study of Jean-Christophe Yoccoz is on the space of quadratic polynomials, written as $p_c \colon z \mapsto z^2 + c$: each (complex) parameter c describes a different dynamical system (coming from iterating the polynomials p_c). Each polynomial p_c has its own filled-in Julia set K_c consisting of those points $z \in \mathbf{C}$ that, under iteration of K_c, have bounded orbits. The *Mandelbrot set* is the set of parameters $c \in \mathbf{C}$ for which the set K_c is connected. The topology of the sets K_c and of the Mandelbrot set is extremely complicated. Yoccoz has provided strong results about the topology of these sets, introducing a tool that is now called the *Yoccoz puzzle* (in technical terms, he proved that many sets K_c are locally connected, and that the Mandelbrot set is locally connected at the corresponding parameters c). The question whether the Mandelbrot set is locally connected everywhere remains one of the big open questions in this field to this day.

Adrien Douady (1935–2006), one of the great French mathematicians of the 20th century, described Yoccoz' work as follows: He combines an extremely acute geometric intuition, an impressive command of analysis, and a penetrating combinatorial sense to play the chess game at which he excels. He occasionally spends half a day on mathematical "experiments", by hand or by computer. "When I make such an experiment", he says, "it is not just the results that interest me, but the manner in which it unfolds, which sheds light on what is really going on."

Small Divisors: Number Theory in Dynamical Systems

Jean-Christophe Yoccoz

1 Planetary Systems

Celestial mechanics is the study of the motion of celestial bodies under Newton's law of gravitation. This law stipulates that the attractive force between any two bodies (which we assume to have negligible size) is proportional to their masses and inversely proportional to the square of their distance. The acceleration of each body is proportional to the resultant of the forces to which it is submitted. In mathematical terms, the motion of N bodies in gravitational interaction is therefore the solution of a system of differential equations of second order.

When there are only 2 bodies, this system can actually be solved explicitly and leads to the famous Kepler laws, which were discovered experimentally long before Newton's law was established: the bodies either escape to infinity (the uninteresting case) or they move periodically along homothetic elliptic orbits.

When there are at least 3 bodies, the system of differential equations becomes fantastically complicated and remains largely mysterious even today. Poincaré showed at the end of the 19-th century that in some appropriate sense, one cannot write the solutions of the system through explicit formulas (somewhat like Galois's assertion, several decades earlier, about the impossibility to solve by radicals the general polynomial equation of degree 5 or larger). Poincaré then looked for other methods in order to study the solutions, founding the modern theory of dynamical systems [5].

Planetary systems constitute a particularly interesting special case of the general N-body problem. One of the bodies (the sun) is assumed to be much heavier than the others (the planets). Therefore, in a first approximation, one can forget about the gravitational interaction between the planets. Each planet will then, independently of the others, move periodically along an ellipsis with the sun as focus. When the

Jean-Christophe Yoccoz
Collège de France, 3 rue d'Ulm, 75231 Paris Cédex 05, France.
e-mail: `jean-c.yoccoz@college-de-france.fr`

motions of all the planets are considered together, the motion is no longer periodic unless the periods of the planets are all commensurate: such a superposition of periodic motions (with not necessarily commensurate periods) is called *quasiperiodic*.

A major question is to understand how much this picture changes when one takes into account the mutual gravitational attraction between the planets. In the short or medium term (a few revolutions around the sun), the effect will not be very important because the perturbation is so much smaller than the attractive force of the sun. But in the long term, the effect is quite significant, at least when some periods are close to being commensurate. For instance, the period of Jupiter is close to 2/5 of the period of Saturn, and for the orbits of these two planets this produces deviations from the Keplerian solutions that have been documented by astronomers several centuries ago.

The question of stability of quasiperiodic motions under small perturbations has been one of the major areas of research in dynamical systems theory for one century. Negative results appeared in the first decades of the 20-th century. Then in 1942 Siegel achieved the first breakthrough in a setting which is described below. In the setting of mechanics which is appropriate for planetary systems, a number of results have been obtained since the 1950's; these are collectively known as KAM theory after Kolmogorov, Arnold, and Moser who were the pioneers in this line of research. A very good survey is [1].

2 Complex Quadratic Polynomials

In this section, we will consider sequences $(z_n)_{n \geq 0}$ of complex numbers which are defined by their initial term z_0 and some recurrence relation $z_{n+1} = f(z_n)$. The map f is fixed and we want to understand the behaviour of the sequence $(z_n)_{n \geq 0}$ as the integer n (that should be thought of as time) goes to ∞. This in general requires different tools depending on the nature of the transformation f; we will only consider examples related to the stability of quasiperiodic motions.

The reference example of a pure unperturbed quasiperiodic motion is given by

$$z_{n+1} = \lambda z_n ,$$

i.e. $f(z) = \lambda z$. Here λ is a fixed complex number of absolute value 1; such a number can be written uniquely as $\lambda = \exp(2\pi i \alpha)$ where α is a real number in $[0, 1)$. Geometrically, z_{n+1} is obtained from z_n by a rotation of angle $2\pi\alpha$ centered at the origin of the complex plane. For this very simple example, we can find an explicit formula for the full sequence:

$$z_n = \lambda^n z_0 = \exp(2\pi i n \alpha) z_0 .$$

We thus have to distinguish two cases:

- α *is a rational number* $\frac{p}{q}$ (with coprime p and q). In this case, we have $\lambda^q = 1$, hence $z_{n+q} = z_n$ for all $n \geq 0$ and the sequence (z_n) is periodic of period q.
- α *is an irrational number*. Except in the trivial case $z_0 = 0$, the z_n are all distinct and lie on the circle centered at the origin of radius $|z_0|$. It is not difficult to show that the sequence z_n is actually dense on this circle: for every point z on the circle and every $\delta > 0$, there exist infinitely many z_n whose distance to z is smaller than δ.

We will now consider a very specific perturbation of the previous example where the recurrence relation is

$$z_{n+1} = \lambda z_n + z_n^2 \,,$$

i.e., $f(z)$ is the complex quadratic polynomial $\lambda z + z^2$. We will assume that the initial value z_0 is small (in absolute value); for small z, the quadratic term z^2 is much smaller than the linear term λz and the new example is indeed a small perturbation of the previous one.

The case of rational α is quite interesting, but we will limit ourselves to the following simple remark. Assume that $\alpha = 0$, so that $z_{n+1} = z_n + z_n^2$, and that z_0 is real and close to 0. Then the sequence (z_n) is converging to 0 if $z_0 < 0$, and increasing to $+\infty$ if $z_0 > 0$. So the behaviour is completely different from the unperturbed case $z_{n+1} = z_n$.

Assume now that α is irrational. Consider the following question:

(*)   *Is the sequence* (z_n) *bounded when* z_0 *is close enough to* 0?

We have just seen that the answer is no for $\alpha = 0$, and the same holds actually for any rational α. For the unperturbed linear example, the answer obviously is yes for all $\alpha \in [0, 1)$.

It can be shown that the answer to the above question is positive if and only if the following much stronger property holds:

(**)   *In a neighbourhood of the origin there exists a change of variables* $z = h(y)$, *defined by a convergent power series* $h(y) = y + \sum_{\ell \geq 2} h_\ell y^\ell$, *such that after setting* $z_n = h(y_n)$ *the recurrence relation* $z_{n+1} = \lambda z_n + z_n^2$ *is transformed into* $y_{n+1} = \lambda y_n$.

In other terms, h must satisfy the functional equation

(FE) $\lambda h(y) + h(y)^2 = h(\lambda y) \,,$

for y close to the origin.

Property (**) means that the behaviour of the sequences (z_n) is deformed, but not qualitatively changed, by the introduction of the quadratic term z_n^2 in the recurrence relation. On the other hand, if property (**) is not satisfied, property (*) is also not satisfied: there exist arbitrarily small initial values z_0 such that the sequence (z_n) is unbounded.

In the beginning of the 20-th century, Cremer [3] constructed examples of irrational numbers α which do not satisfy property (**). Observe first that

$$z_1 = \lambda z_0 + z_0^2 \,,$$
$$z_2 = \lambda^2 z_0 + (\lambda + \lambda^2) z_0^2 + 2\lambda z_0^3 + z_0^4 \,,$$
$$z_3 = \lambda^3 z_0 + \cdots + z_0^8 \,,$$

and more generally

$$z_n = \lambda^n z_0 + \cdots + z_0^{2^n} =: P_{n,\lambda}(z_0) \,.$$

Therefore the product of the $2^n - 1$ non-zero solutions of $P_{n,\lambda}(z_0) - z_0$ is equal to $1 - \lambda^n$. If z_0^* is a solution, the sequence (z_n) with initial value $z_0 = z_0^*$ is periodic of period n. On the other hand, there exists such a solution satisfying

$$|z_0^*| \leq |\lambda^n - 1|^{\frac{1}{2^n - 1}} \,.$$

Assume that λ satisfies

(Cr) $$\inf_{n \geq 1} |\lambda^n - 1|^{\frac{1}{2^n - 1}} = 0 \,.$$

Then we conclude that there are periodic sequences (z_n) starting arbitrarily close to 0 (but not at 0). This means that property $(**)$ is not satisfied, because the unperturbed reference example does not have such periodic sequences.

It remains to see that there are irrational numbers α that satisfy (Cr). To define such a number, let $b_0 = 2$, $b_{k+1} = b_k^{2^{b_k}}$ for $k \geq 0$, and $\alpha = \Sigma_{k \geq 0} b_k^{-1}$. It is easy to check that $\lambda = \exp(2\pi i \alpha)$ satisfies (Cr).

One can also try to calculate directly the coefficients h_ℓ in property $(**)$. One obtains complicated formulas as ℓ gets large where products of terms of the form $\lambda^j - 1$ appear in the denominator. These terms can be very small, making the series $h(y) = y + \Sigma_{\ell \geq 2} h_\ell y^\ell$ divergent even for small nonzero values of y. This explains the name "small divisors" coined for this problem and others of the same nature.

In 1942, Siegel [6] proved the following remarkable result:

Theorem 7.1 (Siegel). *If λ satisfies the Diophantine Condition*

$$(DC)_{\gamma,\tau} \qquad\qquad |\lambda^n - 1| \geq \frac{\gamma}{n^{1+\tau}}$$

for some constants $\gamma > 0$, $\tau \geq 0$ and for all $n > 0$, then property $(**)$ holds.

Siegel's result holds for perturbations of the reference linear example which are much more general than the quadratic one that we have considered. More precisely, it holds for recurrence relations of the form

$$z_{n+1} = \lambda z_n + \sum_{\ell \geq 2} f_\ell z_n^\ell \,,$$

provided the f_ℓ grow at most exponentially fast: there exists $M > 0$ such that $|f_\ell| \leq M^\ell$ for $\ell \geq 2$. In this case, the series is convergent for $|z_n| < M^{-1}$.

Conditions such as (Cr) and $(DC)_{\gamma,\tau}$ are related to the approximation of irrational numbers by rational numbers, a subject discussed in the next section.

3 Diophantine Approximation

Given an irrational number α and any $\varepsilon > 0$, there exists a rational number $\frac{p}{q}$ such that $|\alpha - \frac{p}{q}| < \varepsilon$. But as ε becomes small, q (and $p \approx \alpha q$) must become large. How fast in terms of ε?

The *continued fraction algorithm* produces, for every irrational number α, a sequence of rational numbers (p_k/q_k) called the *convergents* of α which are, in a sense explained below, the best rational approximations of α. The algorithm also analyzes the quality of these approximations.

We denote by $[x]$ the integral part of a real number x and by $\{x\}$ its fractional part. Given an irrational number α, define $a_0 = [\alpha]$, $\alpha_1 = \{\alpha\}$, and then $a_k = [\alpha_k^{-1}]$, $\alpha_{k+1} = \{\alpha_k^{-1}\}$ for $k \geq 1$. Thus, one obtains recursively

$$\text{(CF)} \qquad \alpha = a_0 + \cfrac{1}{a_1 + \cfrac{1}{a_2 + \cfrac{1}{a_3 + \cdots}}}.$$

For $k \geq 0$, define the k-th *convergent* of α as

$$\frac{p_k}{q_k} = a_0 + \cfrac{1}{a_1 + \cfrac{1}{\cdots + \frac{1}{a_k}}}$$

(in lowest terms). The sequences of integers (p_k), (q_k) satisfy the following recurrence relation:

$$p_k = a_k p_{k-1} + p_{k-2}, \quad q_k = a_k q_{k-1} + q_{k-2}$$

starting with $p_{-2} = q_{-1} = 0$, $p_{-1} = q_{-2} = 1$. For instance, for the *golden mean* $\alpha = \frac{\sqrt{5}+1}{2}$, all a_k are equal to 1 and $(p_k = q_{k+1})$ is the Fibonacci sequence.

Conversely, for *any* sequence (a_k) of integers with $a_k \geq 1$ for $k \geq 1$, the formula (CF) defines a unique irrational number α.

The convergents are the best rational approximations of α in the following sense: let $k \geq 0$ and let p, q be integers with $0 < q < q_{k+1}$; if one has

$$|q\alpha - p| \leq |q_k \alpha - p_k|,$$

then $q = q_k$ and $p = p_k$.

Concerning the quality of the approximation by the convergents, one has for all $k \geq 0$ the following estimates:

$$\frac{1}{(a_{k+1}+2)q_k} \leq \frac{1}{q_{k+1}+q_k} < |q_k\alpha - p_k| < \frac{1}{q_{k+1}} \leq \frac{1}{a_{k+1}q_k}.$$

Therefore, large a_{k+1} correspond to especially good rational approximations of α. The golden mean is thus the irrational number with the worst rational approximations.

From the inequalities above, it is easy to see that, for any $\tau \geq 0$, a number α satisfies property $(DC)_{\gamma,\tau}$ (in the statement of Siegel's theorem) for some $\gamma > 0$ if and only if

$$a_{k+1} = O(q_k^\tau).$$

For instance, the golden mean satisfies $(DC)_{\gamma,0}$ for some appropriate $\gamma > 0$. Actually, for any irrational number α which is the solution of a second-degree equation with integer coefficients (the golden mean satisfies $\alpha^2 = \alpha + 1$), the sequence (a_k) becomes periodic when k is large, hence is bounded, and α satisfies $(DC)_{\gamma,0}$ for some appropriate $\gamma > 0$.

A much deeper and more difficult theorem in this direction is Roth's theorem.

Theorem 7.2 (Roth). *If α is an irrational number which is the root of a polynomial (of any degree) with integer coefficients, then, for any $\tau > 0$, there exists $\gamma > 0$ such that α satisfies $(DC)_{\gamma,\tau}$.*

Assume that you choose at random a number $\alpha \in [0,1)$, by choosing successively and independently the digits in its decimal expansion with equal probability. Then α will be irrational almost surely and the corresponding sequences (a_k), (q_k) will satisfy the following properties with full probability:

- the sequence $\left(\dfrac{a_k}{k \log k}\right)_{k \geq 2}$ is unbounded;

- for any $\varepsilon > 0$, the sequence $\left(\dfrac{a_k}{k(\log k)^{1+\varepsilon}}\right)_{k \geq 2}$ is bounded;

- the sequence $\left(\dfrac{1}{k} \log q_k\right)$ converges to $\dfrac{\pi^2}{12 \log 2}$.

A more general version of the first two assertions is due to Khinchin. See [4].

4 Further Results and Open Questions

Let α be an irrational number and $\lambda = \exp(2\pi i \alpha)$. Siegel's theorem states that if the convergents (p_k/q_k) of α satisfy

(DC) $$q_{k+1} = O(q_k^{1+\tau})$$

for some $\tau \geq 0$ then the sequences defined by $z_{n+1} = \lambda z_n + z_n^2$ satisfy the equivalent properties (*) and (**) above. Observe that a "random" number α satisfies (DC) almost surely.

On the other hand, it is easy to check that Cremer's condition (Cr) above is equivalent to

(Cr)′ $$\sup_{k \geq 0} \frac{\log q_{k+1}}{2^{q_k}} = +\infty.$$

Under this condition, we know that the equivalent properties (*) and (**) are not satisfied.

What about irrational numbers α which satisfy neither (DC) nor (Cr)? There is a rather large gap between the growth of the q_k implied by the two conditions.

In 1965, Brjuno [2] proved that, if the convergents of α satisfy

(Br) $$\sum_{k \geq 0} \frac{\log q_{k+1}}{q_k} < +\infty,$$

then properties (*) and (**) are still satisfied. Observe that the growth of the q_k allowed by (Br) is much less restrictive than for (DC); for instance, the condition $q_{k+1} = O(\exp(\sqrt{q_k}))$ implies (Br).

Brjuno's theorem is valid for the same kind of general recurrence relation $z_{n+1} = \lambda z_n + \sum_{\ell \geq 2} f_\ell z_n^\ell$ as Siegel's theorem.

On the other hand, in 1988 I proved the following result [7]. Assume that Brjuno's condition (Br) is not satisfied:

$$\sum_{k \geq 0} \frac{\log q_{k+1}}{q_k} = +\infty.$$

Then the properties (*), (**) are not satisfied for the sequences defined by the quadratic relation $z_{n+1} = \lambda z_n + z_n^2$. In particular, there exist initial values z_0 arbitrarily close to 0 such that the sequence (z_n) converges to ∞.

For the quadratic recurrence relation $z_{n+1} = \lambda z_n + z_n^2$, we therefore know exactly for which irrational numbers α the equivalent properties (*) and (**) are satisfied.

However, this is not the end of the story. Replace the quadratic recurrence relation by

$$z_{n+1} = \lambda z_n + \sum_{2 \leq \ell \leq d} f_\ell z_n^\ell$$

with $f_d \neq 0$. As before, write $\lambda = \exp(2\pi i \alpha)$. From Brjuno's theorem, we know that if α satisfies (Br), then properties (*) and (**) are satisfied. On the other hand, it is conjectured that if α does not satisfy (Br), then properties (*), (**) are not satisfied; but we do not have a proof at the moment.

5 Several Degrees of Freedom

Finally, we come back to the setting of planetary systems introduced in the first section. Consider a bounded solution of the unperturbed system (i.e., we do not take the mutual interaction between the planets into account). Each of the $N-1$ planets describes a Keplerian elliptic orbit with a period T_i ($1 \leq i \leq N-1$). Let

$\omega_i = T_i^{-1}$, $1 \le i \le N - 1$, be the corresponding frequencies. The *totally irrational* (or *non-resonant*) case occurs when there is no relation of the form

$$(\text{Res}) \qquad \sum_{i=1}^{N-1} k_i \, \omega_i = 0$$

with $k_i \in \mathbb{Z}$, not all 0.

One says that the frequency vector $\omega = (\omega_i)$ is *Diophantine* if there exist constants $\gamma > 0$, $\tau \ge 0$ such that for any nonzero vector $k = (k_i) \in \mathbb{Z}^{N-1}$ one has

$$(\text{HDC})_{\gamma, \tau} \qquad \left| \sum_{i=1}^{N-1} k_i \, \omega_i \right| \ge \gamma \left(\sum_{i=1}^{N-1} |k_i| \right)^{2-N-\tau}.$$

We do not try to give a precise mathematical statement from KAM theory which applies in the setting of planetary systems. The general idea is that those solutions of the unperturbed system whose frequency vector is Diophantine (with $\tau > 0$ fixed, and γ not too small with relation to the size of the perturbation) will survive as slightly deformed quasiperiodic solutions of the perturbed system with the same frequency vector. Because a random frequency vector has a strictly positive probability to verify the required condition $(\text{HDC})_{\gamma, \tau}$ (it is not full probability because γ cannot be too small), it will also be true that a random initial condition for the differential equation of the planetary system leads with strictly positive probability to a quasiperiodic solution with a Diophantine frequency vector.

However, we expect that another set of initial conditions having strictly positive probability leads to solutions which are *not* quasiperiodic. To prove this statement and to understand these solutions is a major open problem.

Recommended Reading

1. Jean-Benoît Bost, *Tores invariants des systèmes dynamiques hamiltoniens (d' après Kolmogorov, Arnol'd, Moser, Rüssmann, Zehnder, Herman, Pöschel, ...)* (French). Seminar Bourbaki, Vol. 1984/85. Astérisque 133–134 (1986), 113–157.
2. Alexander D. Brjuno, *Analytical form of differential equations*. Transactions of the Moscow Mathematical Society **25** (1971), 131–288; **26** (1972), 199–239.
3. Hubert Cremer, *Über die Häufigkeit der Nichtzentren* (German). Mathematische Annalen **115** (1938), 573–580.
4. Serge Lang, *Introduction to Diophantine approximations. Second edition.* Springer-Verlag, New York, 1995.
5. Henri Poincaré, *Les méthodes nouvelles de la mécanique céleste. Tome I. Solutions périodiques. Non-existence des intégrales uniformes. Solutions asymptotiques. Tome II. Méthodes de MM. Newcomb, Gyldén, Lindstedt et Bohlin. Tome III. Invariants intégraux. Solutions périodiques du deuxième genre. Solutions doublement asymptotiques* (French). First published in 1892–1899; Dover Publications, Inc., New York, 1957.
6. Carl L. Siegel, *Iteration of analytic functions*. Annals of Mathematics (2) **43** (1942), 607–612.
7. Jean-Christophe Yoccoz, *Théorème de Siegel, nombres de Bruno et polynômes quadratiques. Petits diviseurs en dimension 1* (French). Astérisque 231 (1995), 3–88.

Part II
History: 50 Years International Mathematical Olympiads

Chapter 8
Brief Survey

The International Mathematical Olympiad (IMO) started in 1959 when Romania invited six other socialist countries to a one-off event. The seven participating countries at the 1st IMO were: Bulgaria, Czechoslovakia, German Democratic Republic, Hungary, Poland, Romania and the Union of the Soviet Socialist Republics. A year later Romania repeated the competition, but only 5 countries took part. The next hosts, Hungary, Czechoslovakia, and Poland, made the IMO an annual competition. The number of participating countries increased steadily from year to year up to the current record of 104 countries at the 50th IMO 2009 in Germany - the first time the 100 mark was exceeded. Several breakthroughs played an important role in this development: for example in 1964 Mongolia was the first non-European country and in 1965 Finland was the first non-socialist country to take part. In 1967 France, Italy, Sweden and the United Kingdom joined the IMO competition. In 1974 a team from the United States of America participated for the first time. Three years later the Federal Republic of Germany, the host of the 50th IMO 2009, followed. At the end of the 1970s, when about 20 countries, half socialist and half non-socialist, were taking part, some problems emerged. As a result, three countries who had traditionally taken part, missed the IMO in 1978, namely the German Democratic Republic, Hungary and the Union of the Soviet Socialist Republics. And more dramatically, in 1980 there was no Olympiad at all. It was said that Mongolia had initially wanted to host the 1980 IMO, but had then declined to do so. It certainly seemed as if the future of the IMO was in danger. Fortunately Hungary and France organized the 1982 and 1983 IMOs respectively at very short notice. This was important for the survival of the IMO, even though they had to reduce the number of students allowed in each team. From the 1st IMO in 1959 the standard size of a team was 8, in 1982 it was 4 and since 1983 to today it has been 6. Moreover, supported by ICME, a body was created with the responsibility of finding and organising future IMO hosts. It was first called the "IMO Site Committee" and from 1993 the "IMO Advisory Board". The Advisory Board has 9 members, of which 5 are elected and 4 represent the IMO countries responsible for the Olympiad before and after the year in question.

In the 1980s the number of countries doubled and at the 30th IMO in 1989 the 50 mark was hit for the very first time. In the 1990s there was another big increase when the Soviet Union collapsed and almost all the new states participated in their own right. A similar development occurred with Yugoslavia. In 1997 the Olympiad in Argentina reached a new record of 82 participants. This constant growth continued and 2009 in Germany 104 took part.

The tables in Chapter 9 to 11 show the history of the IMO. The official IMO country codes are used throughout (see Appendix A). Chapter 9 gives a complete list of all 50 IMOs. In addition to the number, the year, the host country and city, we show the number of participating countries (C) and the number of students (P). Moreover, (S) shows as a percentage the points obtained by all competitors, which provides an index of the level of difficulty of each Olympiad. By this measure the easiest IMO was in 1966, the hardest one in 1971. In recent years many IMOs were fairly difficult with around 30-35% only.

Although the IMO is a competition for individuals only, it is interesting to see which countries have the highest number of points of its participants. This provides some measure of the difficulty of each IMO for the top teams. (TC) shows the best team and (TS) its percentage points for its students. By this measure the 1994 IMO was the easiest one, since this was the only time when a full team got 100%. The hardest IMO was the 1977 one, where only 63% of the points were given to the winning team.

In total, **12911** students have taken part in the 50 IMOs. They were awarded the following number of prizes.

1056	gold medals	8.2%
2135	silver medals	16.5%
3191	bronze medals	24.7%
6382	total	49.4%

In Chapter 10, we list the number of IMOs in which each country has taken part (N), as well as the number of problems (AP) they have proposed. Curiously, in 1994 Armenia and Australia proposed the same problem; we count this as 0.5 problems. (H) displays the number of IMOs hosted by each country. (P) shows the number of participants, the medals won by the teams are shown in (G) gold, (S) silver and (B) bronze. At the first 3 IMOs diplomas were presented as "fourth" prizes, but we do not list these. From the very beginning there have also been special prizes for outstanding solutions to single problems as shown in (SP). In recent years these prizes have been extremely rare. Since 1988 there have been "Honourable mentions" (HM) for those students who obtained the full score of 7 points for a particular problem.

In Chapter 11, we list the complete results for all teams and all IMOs.

The IMO regulations have developed over time, but the basic rules have remained constant, such as having two competition papers of about 4½ hours each on two consecutive days. In general, each paper consists of 3 problems, i.e. 6 problems for each IMO. There have only been two exceptions, namely the second and fourth

IMOs had 7 problems each. Hence the total number of problems at all 50 IMOs is 302. In addition, the number of points awarded for each problem varied from 40 to 46 during the first 20 years. Since 1981 each problem has been worth 7 points, with a maximum of 42 points in total.

The score levels for the gold, silver and bronze medals have prizes varied as well. In the early years natural gaps were used. For example, in 1969 there were 3 full scores of 40 points, but no participants scored 39 or 38 points. The jury decided to award 3 gold medals only, i.e. less than 3%. This was the IMO with the lowest percentage of gold medals. About 20 years ago the regulations made the rules consistent: no more than 50% of participants should get a medal and the ratio of gold, silver and bronze medals should be close to 1:2:3. Therefore, about $1/12$ of the students will get a gold, $1/6$ silver and $1/4$ bronze medals. In general this rule determines the score levels immediately. But sometimes very unusual score distributions have led to strange situations. The most unusual case was in 2006 in Slovenia when 65 out of 498 students got 15 points. A cut at 15 points would yield 50.8% of medals, whereas 16 points would result in 37.8%, i.e. around $1/3$ of the students only. The jury decided to be generous - certainly the best decision in the spirit of IMO events.

Chapter 9
All IMOs

For the IMO country codes, see the listing in Appendix A. Besides the year, we list the corresponding IMO's host country and city. The column title abbreviations refer to the following.

C   number of participating countries
P   number of participating students
S   overall scores (percentage of maximum scores)
TC   top team country
TS   top team scores (percentage of maximum scores)

No.	Year	Host	City	C	P	S	TC	TS
1	1959	ROU	Braşov	7	52	52	ROU	78
2	1960	ROU	Sinaia	5	39	56	CZS	71
3	1961	HUN	Veszprem	6	48	56	HUN	84
4	1962	CZS	České Budějovice	7	56	61	HUN	79
5	1963	POL	Wroclaw	8	64	56	USS	85
6	1964	USS	Moscow	9	72	61	USS	80
7	1965	GDR	Berlin	10	80	50	USS	88
8	1966	BGR	Sofia	9	72	74	USS	92
9	1967	YUG	Cetinje	13	99	52	USS	82
10	1968	USS	Moscow	12	96	71	GDR	95
11	1969	ROU	Bucharest	14	112	50	HUN	77
12	1970	HUN	Keszthely	14	112	48	HUN	73
13	1971	CZS	Žilina	15	115	28	HUN	80
14	1972	POL	Toruń	14	107	47	USS	84
15	1973	USS	Moscow	16	125	44	USS	79
16	1974	GDR	Erfurt	18	140	56	USS	80

No.	Year	Host	City	C	P	S	TC	TS
17	1975	BGR	Burgas	17	135	55	HUN	81
18	1976	AUT	Lienz	18	139	43	USS	78
19	1977	YUG	Belgrade	21	155	44	USA	63
20	1978	ROU	Bucharest	17	132	51	ROU	74
21	1979	UNK	London	23	166	49	USS	83
22	1981	USA	Washington	27	185	63	USA	93
23	1982	HUN	Budapest	30	119	49	GER	86
24	1983	FRA	Paris	32	186	37	GER	84
25	1984	CZS	Prague	34	192	42	USS	93
26	1985	FIN	Joutsa	38	209	35	ROU	80
27	1986	POL	Warsaw	37	210	43	USA, USS	81
28	1987	CUB	Havana	42	237	47	ROU	99
29	1988	AUS	Canberra	49	268	36	USS	86
30	1989	GER	Braunschweig	50	291	45	CHN	94
31	1990	CHN	Beijing	54	308	41	CHN	91
32	1991	SWE	Sigtuna	56	318	46	USS	96
33	1992	RUS	Moscow	56	322	36	CHN	95
34	1993	TUR	Istanbul	73	413	30	CHN	85
35	1994	HKG	Hong Kong	69	385	48	USA	100
36	1995	CAN	Toronto	73	412	45	CHN	94
37	1996	IND	Mumbai	75	424	30	ROU	74
38	1997	ARG	Mar del Plata	82	460	38	CHN	88
39	1998	TWN	Taipeh	76	419	35	IRN	84
40	1999	ROU	Bucharest	81	450	32	CHN, RUS	72
41	2000	KOR	Taejon	82	461	32	CHN	87
42	2001	USA	Washington	83	473	31	CHN	89
43	2002	UNK	Glasgow	84	479	33	CHN	84
44	2003	JPN	Tokyo	82	457	31	BGR	90
45	2004	HEL	Athens	85	486	39	CHN	87
46	2005	MEX	Mérida	91	513	33	CHN	93
47	2006	SVN	Ljubljana	90	498	34	CHN	85
48	2007	VNM	Hanoi	93	520	33	RUS	73
49	2008	ESP	Madrid	97	535	36	CHN	86
50	2009	GER	Bremen	104	565	36	CHN	88

Chapter 10
All Countries

For the IMO country codes, see the listing in Appendix A. Every country's first participation is shown in column *Year*. The column title abbreviations refer to the following numbers.

N   IMOs in which the country took part
AP   problems proposed by the country
H   IMOs that were hosted by the country
P   participants
G   Gold medals
S   Silver medals
B   Bronze medals
SP   Special prices for outstanding solutions
HM   Honorable mentions

	Country	Year	N	AP	H	P	G	S	B	SP	HM
1	ALB	1993	14	-	-	74	-	2	5	-	14
2	ALG	1977	13	-	-	62	-	1	2	-	2
3	ARG	1988	21	-	1	123	3	19	45	-	16
4	ARM	1993	17	1.5	-	100	1	9	35	-	23
5	AUS	1981	29	4.5	1	174	13	44	69	-	14
6	AUT	1970	39	1	1	254	12	27	84	-	31
7	AZE	1993	16	-	-	88	-	4	18	-	20
8	BAH	1990	3	-	-	18	-	-	-	-	1
9	BEL	1969	31	3	-	191	1	10	46	-	34
10	BEN	2009	1	-	-	2	-	-	-	-	-
11	BGD	2005	5	-	-	25	-	-	2	-	9
12	BGR	1959	50	16	2	342	51	92	90	4	1
13	BIH	1993	17	-	-	94	-	4	25	-	24

	Country	Year	N	AP	H	P	G	S	B	SP	HM
14	BLR	1993	17	4	-	100	12	36	40	-	5
15	BOL	1997	6	-	-	17	-	-	-	-	1
16	BRA	1979	30	1	-	177	8	21	58	-	22
17	BRU	2000	1	-	-	2	-	-	-	-	-
18	CAN	1981	29	2	1	174	17	40	68	-	16
19	CHI	1994	7	-	-	23	-	3	1	-	5
20	CHN	1985	24	3	1	140	107	26	5	-	-
21	CIS	1992	1	-	-	6	2	3	-	-	1
22	COL	1981	29	1	-	174	1	15	53	-	23
23	CRI	2005	5	-	-	23	-	-	5	-	8
24	CUB	1971	36	1	1	155	1	6	36	-	21
25	CYP	1984	25	-	-	140	-	2	11	-	21
26	CZE	1993	17	3	-	102	3	22	42	-	18
27	CZS	1959	33	16	3	237	10	50	73	1	2
28	DEN	1991	19	-	-	108	-	4	19	-	21
29	ECU	1988	11	-	-	61	-	-	3	-	8
30	ESP	1983	27	-	1	156	-	3	28	-	33
31	EST	1993	17	2	-	99	-	4	19	-	25
32	FIN	1965	36	7	1	230	1	5	47	1	32
33	FRA	1967	40	8	1	259	23	42	85	-	20
34	GDR	1959	29	9	2	210	26	62	60	4	-
35	GEO	1993	17	-	-	102	2	12	41	-	30
36	GER	1977	32	13	2	198	47	76	55	-	8
37	GTM	1997	8	-	-	35	-	-	1	-	2
38	HEL	1975	31	1	1	192	-	15	51	-	37
39	HKG	1988	22	-	1	132	4	34	57	-	15
40	HND	2008	2	-	-	5	-	-	1	-	2
41	HRV	1993	17	-	-	102	-	6	44	-	26
42	HUN	1959	49	15	3	330	75	141	79	8	4
43	IDN	1988	21	-	-	122	-	3	16	-	26
44	IND	1989	21	4	1	126	8	52	48	-	12
45	IRL	1988	22	4	-	132	-	1	7	-	19
46	IRN	1985	24	2	-	139	31	67	28	-	3
47	ISL	1985	25	2	-	125	-	1	9	-	17
48	ISR	1979	28	2	-	167	10	32	76	-	17
49	ITA	1967	30	2	-	173	7	16	58	-	25
50	JPN	1990	20	1	1	120	28	52	31	-	3

	Country	Year	N	AP	H	P	G	S	B	SP	HM
51	KAZ	1993	16	-	-	96	8	17	37	-	18
52	KGZ	1993	17	-	-	87	-	-	7	-	16
53	KHM	2007	3	-	-	16	-	-	-	-	4
54	KOR	1988	22	8	1	132	38	54	25	-	6
55	KWT	1982	25	-	-	113	-	-	1	-	1
56	LIE	2005	5	-	-	10	-	-	2	-	1
57	LKA	1995	14	-	-	59	-	-	8	-	13
58	LTU	1993	17	2	-	102	1	6	21	-	30
59	LUX	1979	24	3	-	55	2	5	18	-	11
60	LVA	1993	17	-	-	102	1	9	30	-	27
61	MAC	1990	20	-	-	119	-	2	16	-	23
62	MAR	1983	27	-	-	161	-	3	28	-	45
63	MAS	1995	15	-	-	76	-	2	5	-	15
64	MDA	1993	17	-	-	92	5	14	31	1	12
65	MEX	1981	24	-	1	142	1	6	37	-	28
66	MKD	1993	17	1	-	98	-	4	37	-	17
67	MNE	2007	3	-	-	10	-	-	-	-	4
68	MNG	1964	38	3	-	254	1	19	40	1	29
69	MOZ	2004	3	-	-	11	-	-	-	-	-
70	MRT	2009	1	-	-	6	-	-	-	-	-
71	NCY	1993	1	-	-	6	-	-	-	-	-
72	NGA	2006	3	-	-	18	-	-	-	-	1
73	NIC	1987	1	-	-	6	-	-	-	-	-
74	NLD	1969	39	22	-	256	2	22	49	-	34
75	NOR	1984	26	-	-	148	2	10	27	-	18
76	NZL	1988	22	3	-	132	1	4	33	-	28
77	PAK	2005	4	-	-	22	-	-	2	-	2
78	PAN	1987	3	-	-	11	-	-	-	-	3
79	PAR	1997	11	-	-	44	-	1	1	-	6
80	PER	1987	16	-	-	81	1	11	25	-	23
81	PHI	1988	21	1	-	104	-	1	8	-	11
82	POL	1959	49	24	3	328	21	64	112	4	18
83	POR	1989	21	-	-	125	-	1	11	-	14
84	PRI	1997	10	-	-	50	-	1	1	-	2
85	PRK	1990	6	-	-	36	6	14	6	-	1
86	ROU	1959	50	22	5	338	68	113	90	1	2
87	RUS	1992	18	12	1	108	70	29	9	-	-

	Country	Year	N	AP	H	P	G	S	B	SP	HM
88	SAF	1992	18	-	-	108	1	8	28	-	31
89	SAU	2004	5	-	-	25	-	-	-	-	-
90	SCG	2003	3	2	-	18	-	5	7	-	3
91	SGP	1988	22	-	-	132	1	26	60	-	19
92	SLV	2005	5	-	-	20	-	-	-	-	10
93	SRB	2006	4	-	-	24	3	6	10	-	3
94	SUI	1991	19	-	-	95	1	8	22	-	23
95	SVK	1993	17	-	-	102	3	25	45	-	15
96	SVN	1993	17	-	1	99	-	3	24	-	26
97	SWE	1967	42	7	1	277	5	23	68	1	30
98	SYR	2009	1	-	-	5	-	-	-	-	-
99	THA	1989	21	-	-	126	6	27	39	-	21
100	TJK	2005	5	-	-	27	-	1	4	-	8
101	TKM	1993	12	-	-	56	-	2	13	-	13
102	TTO	1991	19	-	-	111	-	-	4	-	20
103	TUN	1981	18	-	-	84	1	2	12	-	6
104	TUR	1978	26	1	1	158	10	37	59	-	11
105	TWN	1992	18	1	1	108	23	63	17	-	4
106	UAE	2008	2	-	-	9	-	-	-	-	-
107	UKR	1993	17	2	-	102	26	37	27	-	6
108	UNK	1967	42	22	2	278	35	85	108	9	11
109	URY	1987	13	-	-	55	-	-	1	-	8
110	USA	1974	35	10	2	222	82	100	29	2	1
111	USS	1959	29	20	3	204	77	67	45	2	-
112	UZB	1997	11	-	-	63	-	4	20	-	19
113	VEN	1981	16	-	-	53	-	2	2	-	9
114	VNM	1974	33	3	1	198	44	82	57	-	1
115	YUG	1963	37	4	2	256	6	46	96	1	7
116	ZWE	2009	1	-	-	2	-	-	-	-	-

Chapter 11
All Results

For the IMO country codes, see the listing in Appendix A. The column title abbreviations refer to the following numbers for the country's team.

P   number of contestants
TS   total scores
G   Gold medals
S   Silver medals
B   Bronze medals
H   Honorable mentions

1st IMO 1959 - Romania

	Country	P	TS	G	S	B		Country	P	TS	G	S	B
1	ROU	8	249	1	2	2	5	POL	8	122	-	-	-
2	HUN	8	233	1	1	2	6	USS	4	111	-	-	1
3	CZS	8	192	1	-	-	7	GDR	8	40	-	-	-
4	BGR	8	131	-	-	-							

2nd IMO 1960 - Romania

	Country	P	TS	G	S	B		Country	P	TS	G	S	B
1	CZS	8	257	1	1	2	4	BGR	8	175	-	-	1
2	HUN	8	248	2	2	-	5	GDR	7	38	-	-	-
	ROU	8	248	1	1	1							

3rd IMO 1961 - Hungary

	Country	P	TS	G	S	B		Country	P	TS	G	S	B
1	HUN	8	270	2	3	1	4	CZS	8	159	-	-	1
2	POL	8	203	1	-	-	5	GDR	8	146	-	-	1
3	ROU	8	197	-	1	1	6	BGR	8	108	-	-	-

4th IMO 1962 - Czechoslovakia

	Country	P	TS	G	S	B		Country	P	TS	G	S	B
1	HUN	8	289	2	3	2		POL	8	212	-	1	3
2	USS	8	263	2	2	2	6	BGR	8	196	-	1	2
3	ROU	8	257	-	3	3	7	GDR	8	153	-	1	-
4	CZS	8	212	-	1	3							

5th IMO 1963 - Poland

	Country	P	TS	G	S	B		Country	P	TS	G	S	B
1	USS	8	271	4	3	1	5	CZS	8	151	1	-	1
2	HUN	8	234	-	5	3	6	BGR	8	145	-	-	3
3	ROU	8	191	1	1	3	7	GDR	8	140	-	-	3
4	YUG	8	162	1	2	1	8	POL	8	134	-	-	2

6th IMO 1964 - Union of the Soviet Socialist Republics

	Country	P	TS	G	S	B		Country	P	TS	G	S	B
1	USS	8	269	3	1	3	6	GDR	8	196	-	1	2
2	HUN	8	253	3	1	1	7	CZS	8	194	-	2	2
3	ROU	8	213	-	2	3	8	MNG	8	169	-	-	1
4	POL	8	209	1	1	3	9	YUG	8	155	-	1	1
5	BGR	8	198	-	-	3							

7th IMO 1965 - German Democratic Republic

	Country	P	TS	G	S	B		Country	P	TS	G	S	B
1	USS	8	281	5	2	-	6	CZS	8	159	-	1	3
2	HUN	8	244	3	2	2	7	YUG	8	137	-	-	2
3	ROU	8	222	-	4	3	8	BGR	8	93	-	-	1
4	POL	8	178	-	1	3	9	MNG	8	63	-	-	-
5	GDR	8	175	-	2	3	10	FIN	8	62	-	-	-

8th IMO 1966 - Bulgaria

	Country	P	TS	G	S	B		Country	P	TS	G	S	B
1	USS	8	293	5	1	1	6	BGR	8	236	-	1	3
2	HUN	8	281	3	2	1	7	YUG	8	224	-	2	1
3	GDR	8	280	3	3	-	8	CZS	8	215	-	1	2
4	POL	8	269	1	4	1	9	MNG	8	90	-	-	-
5	ROU	8	257	1	2	2							

9th IMO 1967 - Yugoslavia

	Country	P	TS	G	S	B		Country	P	TS	G	S	B
1	USS	8	275	3	3	2	8	YUG	8	136	-	-	3
2	GDR	8	257	3	3	1	9	SWE	8	135	-	-	2
3	HUN	8	251	2	3	3	10	ITA	6	110	-	1	1
4	UNK	8	231	1	2	4	11	POL	8	101	-	-	1
5	ROU	8	214	1	1	4	12	MNG	8	87	-	-	1
6	BGR	8	159	1	-	1	13	FRA	5	41	-	-	-
	CZS	8	159	-	1	3							

10th IMO 1968 - Union of the Soviet Socialist Republics

	Country	P	TS	G	S	B		Country	P	TS	G	S	B
1	GDR	8	304	5	3	-	7	CZS	8	248	2	4	-
2	USS	8	298	5	1	2	8	ROU	8	208	1	1	2
3	HUN	8	291	3	3	2	9	BGR	8	204	-	3	1
4	UNK	8	263	3	2	2	10	YUG	8	177	-	-	3
5	POL	8	262	2	3	2	11	ITA	8	132	-	-	1
6	SWE	8	256	1	2	5	12	MNG	8	74	-	-	-

11th IMO 1969 - Romania

	Country	P	TS	G	S	B		Country	P	TS	G	S	B
1	HUN	8	247	1	4	2	8	CZS	8	170	-	-	3
2	GDR	8	240	-	4	4	9	MNG	8	120	-	-	1
3	USS	8	231	1	3	3	10	FRA	8	119	-	1	-
4	ROU	8	219	-	4	2		POL	8	119	-	1	-
5	UNK	8	193	1	1	1	12	SWE	8	104	-	-	-
6	BGR	8	189	-	-	3	13	BEL	8	57	-	-	-
7	YUG	8	181	-	2	2	14	NLD	8	51	-	-	-

12th IMO 1970 - Hungary

	Country	P	TS	G	S	B			Country	P	TS	G	S	B
1	HUN	8	233	3	1	3			CZS	8	145	-	-	4
2	GDR	8	221	1	2	4		9	FRA	8	141	-	1	4
	USS	8	221	2	1	3		10	SWE	8	110	-	-	2
4	YUG	8	209	-	3	3		11	POL	8	105	-	-	1
5	ROU	8	208	-	3	4		12	AUT	8	104	-	-	1
6	UNK	8	180	1	-	6		13	NLD	8	87	-	-	1
7	BGR	8	145	-	-	3		14	MNG	8	58	-	-	1

13th IMO 1971 - Czechoslovakia

	Country	P	TS	G	S	B			Country	P	TS	G	S	B
1	HUN	8	255	4	4	-		9	CZS	8	55	-	-	1
2	USS	8	205	1	5	2		10	NLD	8	48	-	-	2
3	GDR	8	142	1	1	4		11	SWE	7	43	-	-	2
4	POL	8	118	1	-	4		12	BGR	8	39	-	-	-
5	ROU	8	110	-	1	4		13	FRA	8	38	-	-	-
	UNK	8	110	-	1	4		14	MNG	8	26	-	-	-
7	AUT	8	82	-	-	4		15	CUB	4	9	-	-	-
8	YUG	8	71	-	-	2								

14th IMO 1972 - Poland

	Country	P	TS	G	S	B			Country	P	TS	G	S	B
1	USS	8	270	2	4	2			YUG	8	136	-	-	3
2	HUN	8	263	3	3	2		9	CZS	8	131	-	-	4
3	GDR	8	239	1	3	4		10	BGR	8	120	-	-	2
4	ROU	8	208	1	3	1		11	SWE	8	60	-	-	2
5	UNK	8	179	-	2	4		12	NLD	8	51	-	-	-
6	POL	8	160	1	1	1		13	MNG	8	48	-	-	-
7	AUT	8	136	-	-	5		14	CUB	3	14	-	-	-

15th IMO 1973 - Union of the Soviet Socialist Republics

	Country	P	TS	G	S	B			Country	P	TS	G	S	B
1	USS	8	254	3	2	3		9	ROU	8	141	-	1	3
2	HUN	8	215	1	2	5		10	YUG	8	137	-	-	5
3	GDR	8	188	-	3	4		11	SWE	8	99	-	1	1
4	POL	8	174	-	2	4		12	BGR	8	96	-	-	1
5	UNK	8	164	1	-	5			NLD	8	96	-	-	2
6	FRA	8	153	-	3	1		14	FIN	8	86	-	-	2
7	CZS	8	149	-	1	4		15	MNG	8	65	-	-	1
8	AUT	8	144	-	-	6		16	CUB	5	42	-	-	1

16th IMO 1974 - German Democratic Republic

	Country	P	TS	G	S	B		Country	P	TS	G	S	B
1	USS	8	256	2	3	2	10	SWE	8	187	1	1	-
2	USA	8	243	-	5	3	11	BGR	8	171	-	1	4
3	HUN	8	237	1	3	3	12	CZS	8	158	-	-	2
4	GDR	8	236	-	5	2	13	VNM	5	146	1	1	2
5	YUG	8	216	2	1	2	14	POL	8	138	-	-	2
6	AUT	8	212	1	1	4	15	NLD	8	112	-	-	1
7	ROU	8	199	1	1	3	16	FIN	8	111	-	-	1
8	FRA	8	194	1	1	3	17	CUB	7	65	-	-	-
9	UNK	8	188	-	1	3	18	MNG	8	60	-	-	-

17th IMO 1975 - Bulgaria

	Country	P	TS	G	S	B		Country	P	TS	G	S	B
1	HUN	8	258	-	5	3	10	VNM	7	175	-	1	3
2	GDR	8	249	-	4	4	11	YUG	8	163	-	1	1
3	USA	8	247	3	1	3	12	CZS	8	162	-	-	2
4	USS	8	246	1	3	4	13	SWE	8	160	-	2	-
5	UNK	8	239	2	2	3	14	POL	8	124	-	1	1
6	AUT	8	192	1	1	2	15	HEL	8	95	-	1	-
7	BGR	8	186	-	1	4	16	MNG	8	75	-	-	1
8	ROU	8	180	-	1	3	17	NLD	8	67	-	-	1
9	FRA	8	176	1	1	1							

18th IMO 1976 - Austria

	Country	P	TS	G	S	B		Country	P	TS	G	S	B
1	USS	8	250	4	3	1	10	SWE	8	120	-	1	3
2	UNK	8	214	2	4	1	11	ROU	8	118	-	1	3
3	USA	8	188	1	4	1	12	CZS	8	116	-	1	3
4	BGR	8	174	-	2	6		YUG	8	116	-	1	3
5	AUT	8	167	1	2	5	14	VNM	8	112	-	1	3
6	FRA	8	165	1	3	1	15	NLD	8	78	-	-	1
7	HUN	8	160	-	3	4	16	FIN	8	52	-	-	1
8	GDR	8	142	-	2	3	17	HEL	8	50	-	-	-
9	POL	8	138	-	-	6	18	CUB	3	16	-	-	-

19th IMO 1977 - Yugoslavia

	Country	P	TS	G	S	B		Country	P	TS	G	S	B
1	USA	8	202	2	3	1	12	AUT	8	151	1	1	2
2	USS	8	192	1	2	4	13	SWE	8	137	1	1	2
3	HUN	8	190	1	3	2	14	FRA	8	126	1	-	-
	UNK	8	190	1	3	3	15	ROU	8	122	-	1	2
5	NLD	8	185	1	2	3	16	FIN	8	88	-	-	1
6	BGR	8	172	-	3	3	17	MNG	8	49	-	-	-
7	GER	8	165	1	1	4	18	CUB	4	41	-	-	-
8	GDR	8	163	2	1	1	19	BEL	7	33	-	-	-
9	CZS	8	161	-	3	2	20	ITA	5	22	-	-	-
10	YUG	8	159	-	3	3	21	ALG	3	17	-	-	-
11	POL	8	157	1	2	2							

20th IMO 1978 - Romania

	Country	P	TS	G	S	B		Country	P	TS	G	S	B
1	ROU	8	237	2	3	2	10	YUG	8	171	-	1	2
2	USA	8	225	1	3	3	11	NLD	8	157	-	1	1
3	UNK	8	201	1	2	2	12	POL	8	156	-	-	2
4	VNM	8	200	-	2	6	13	FIN	8	118	-	-	2
5	CZS	8	195	-	2	3	14	SWE	8	117	-	-	1
6	GER	8	184	1	-	3	15	CUB	4	68	-	-	2
7	BGR	8	182	-	1	3	16	TUR	8	66	-	-	-
8	FRA	8	179	-	2	4	17	MNG	8	61	-	-	-
9	AUT	8	174	-	3	2							

21st IMO 1979 - United Kingdom

	Country	P	TS	G	S	B		Country	P	TS	G	S	B
1	USS	8	267	2	4	1	13	BGR	8	150	-	-	5
2	ROU	8	240	1	4	2	14	SWE	8	143	-	2	1
3	GER	8	235	1	5	1	15	VNM	4	134	1	3	-
4	UNK	8	218	-	4	4	16	NLD	8	130	-	1	1
5	USA	8	199	1	2	2	17	ISR	8	119	-	-	2
6	GDR	8	180	-	2	2	18	FIN	8	89	-	-	1
7	CZS	8	178	1	-	4	19	BEL	8	66	-	-	1
8	HUN	8	176	-	2	2	20	HEL	8	57	-	-	1
9	YUG	8	168	-	1	4	21	CUB	4	35	-	-	-
10	POL	8	160	-	2	3	22	BRA	5	19	-	-	-
11	FRA	8	155	1	-	1	23	LUX	1	7	-	-	-
12	AUT	8	152	-	-	4							

22nd IMO 1981 - United States of America

	Country	P	TS	G	S	B		Country	P	TS	G	S	B
1	USA	8	314	4	3	1	15	ISR	6	175	1	-	3
2	GER	8	312	5	2	1	16	BRA	8	172	1	-	-
3	UNK	8	301	3	4	1	17	HUN	4	164	3	1	-
4	AUT	8	290	4	2	1	18	CUB	8	141	-	1	-
5	BGR	8	287	2	3	3	19	BEL	8	139	-	2	-
6	POL	8	259	2	3	1	20	ROU	4	136	-	2	2
7	CAN	8	249	2	2	1	21	AUS	8	122	-	-	1
8	YUG	8	246	1	2	3	22	HEL	8	104	-	-	-
9	USS	6	230	3	2	1	23	COL	8	93	-	-	-
10	NLD	8	219	-	3	1	24	VEN	8	64	-	-	-
11	FRA	8	209	2	-	3	25	LUX	1	42	1	-	-
12	SWE	8	207	-	1	3	26	TUN	2	32	-	-	-
13	FIN	8	206	1	1	3	27	MEX	5	12	-	-	-
14	CZS	5	190	1	3	1							

23rd IMO 1982 - Hungary

	Country	P	TS	G	S	B		Country	P	TS	G	S	B
1	GER	4	145	2	2	-	16	AUT	4	82	1	-	1
2	USS	4	137	2	1	1	17	CAN	4	78	-	-	2
3	GDR	4	136	2	1	1	18	ISR	4	75	-	-	1
	USA	4	136	1	2	1	19	SWE	4	74	-	-	2
5	VNM	4	133	1	2	1	20	AUS	4	66	-	-	1
6	HUN	4	125	-	3	1		BRA	4	66	-	-	1
7	CZS	4	115	-	2	2	22	MNG	4	56	-	-	1
8	FIN	4	113	-	2	1	23	HEL	4	55	-	-	-
9	BGR	4	108	-	-	4	24	BEL	4	50	-	-	1
10	UNK	4	103	-	-	4	25	CUB	4	44	-	-	-
11	ROU	4	99	-	1	2	26	COL	4	34	-	-	-
12	YUG	4	98	-	2	-	27	ALG	3	23	-	-	-
13	POL	4	96	-	1	2		VEN	4	23	-	-	-
14	NLD	4	92	-	1	1	29	TUN	4	19	-	-	-
15	FRA	4	89	1	-	-	30	KWT	4	4	-	-	-

24th IMO 1983 - France

	Country	P	TS	G	S	B		Country	P	TS	G	S	B
1	GER	6	212	4	1	-	17	HEL	6	96	-	-	3
2	USA	6	171	1	3	2	18	YUG	6	89	-	-	5
3	HUN	6	170	-	4	2	19	AUS	6	86	-	1	2
4	USS	6	169	1	3	2	20	BRA	6	77	-	-	3
5	ROU	6	161	1	2	3	21	SWE	6	47	-	-	-
6	VNM	6	148	-	3	3	22	AUT	6	45	-	-	-
7	NLD	6	143	1	3	-	23	ESP	4	37	-	-	-
8	CZS	6	142	1	1	3	24	CUB	6	36	-	-	1
9	BGR	6	137	-	1	4	25	MAR	6	32	-	-	-
10	FRA	6	123	-	2	3	26	BEL	6	31	-	-	-
11	UNK	6	121	-	3	1	27	TUN	6	26	-	-	-
12	GDR	6	117	-	-	5	28	COL	6	21	-	-	-
13	FIN	6	103	-	-	3	29	LUX	2	13	-	-	-
14	CAN	6	102	-	-	4	30	ALG	6	6	-	-	-
15	POL	6	101	-	-	3	31	KWT	6	4	-	-	-
16	ISR	6	97	-	-	5	32	ITA	6	2	-	-	-

25th IMO 1984 - Czechoslovakia

	Country	P	TS	G	S	B		Country	P	TS	G	S	B
1	USS	6	235	5	1	-	18	BRA	6	92	-	-	3
2	BGR	6	203	2	3	1	19	HEL	6	88	-	1	-
3	ROU	6	199	2	2	2	20	CAN	6	83	-	-	1
4	HUN	6	195	1	4	1	21	COL	6	80	-	-	2
	USA	6	195	1	4	1	22	CUB	6	67	-	-	1
6	UNK	6	169	1	3	1	23	BEL	6	56	-	-	1
7	VNM	6	162	1	2	3		MAR	6	56	-	-	1
8	GDR	6	161	1	2	3	25	SWE	6	53	-	-	-
9	GER	6	150	-	2	4	26	CYP	6	47	-	-	1
10	MNG	6	146	-	3	2	27	ESP	6	43	-	-	-
11	POL	6	140	-	1	5	28	ALG	4	36	-	-	-
12	FRA	6	126	-	2	2	29	FIN	6	31	-	-	-
13	CZS	6	125	-	2	2	30	TUN	6	29	-	-	-
14	YUG	6	105	-	-	4	31	NOR	1	24	-	-	1
15	AUS	6	103	-	1	2	32	LUX	1	22	-	-	1
16	AUT	6	97	-	1	2	33	KWT	6	9	-	-	-
17	NLD	6	93	-	1	2	34	ITA	6	-	-	-	-

26th IMO 1985 - Finland

	Country	P	TS	G	S	B		Country	P	TS	G	S	B
1	ROU	6	201	3	3	-	20	HEL	6	69	-	1	1
2	USA	6	180	2	4	-	21	YUG	6	68	-	-	2
3	HUN	6	168	2	2	2	22	SWE	6	65	-	-	1
4	BGR	6	165	2	3	-	23	MNG	6	62	-	1	-
5	VNM	6	144	1	3	1	24	BEL	6	60	1	-	1
6	USS	6	140	1	2	2		MAR	6	60	-	-	2
7	GER	6	139	1	1	4	26	COL	6	54	-	-	2
8	GDR	6	136	-	3	3		TUR	6	54	-	-	2
9	FRA	6	125	-	2	3	28	TUN	4	46	-	-	2
10	UNK	6	121	-	2	3	29	ALG	6	36	-	-	-
11	AUS	6	117	1	1	2	30	NOR	6	35	-	-	-
12	CAN	6	105	-	1	4	31	IRN	1	28	-	1	-
	CZS	6	105	-	3	1	32	CHN	2	27	-	-	1
14	POL	6	101	-	1	4		CYP	6	27	-	-	1
15	BRA	6	83	-	-	2	34	FIN	6	25	-	-	-
16	ISR	6	81	-	1	-	35	ESP	4	23	-	-	-
17	AUT	6	77	-	-	3	36	ITA	5	20	-	-	-
18	CUB	6	74	-	-	2	37	ISL	2	13	-	-	-
19	NLD	6	72	-	-	1	38	KWT	5	7	-	-	-

27th IMO 1986 - Poland

	Country	P	TS	G	S	B		Country	P	TS	G	S	B
1	USA	6	203	3	3	-	20	YUG	6	84	-	-	2
	USS	6	203	2	4	-	21	ALG	6	80	-	-	2
3	GER	6	196	2	4	-	22	BEL	6	79	-	1	2
4	CHN	6	177	3	1	1	23	ESP	4	78	-	1	2
5	GDR	6	172	1	3	2	24	BRA	6	69	1	-	-
6	ROU	6	171	2	2	1	25	NOR	6	68	-	1	-
7	BGR	6	161	1	3	2	26	HEL	6	63	-	-	2
8	HUN	6	151	1	2	2	27	FIN	6	60	-	-	1
9	CZS	6	149	-	3	3	28	COL	6	58	-	-	-
10	VNM	6	146	1	2	2	29	SWE	6	57	-	-	1
11	UNK	6	141	-	2	3	30	TUR	6	55	-	-	-
12	FRA	6	131	1	1	2	31	MNG	6	54	-	-	-
13	AUT	6	127	-	2	2	32	CYP	6	53	-	1	-
14	ISR	6	119	-	2	2	33	CUB	6	51	-	-	-
15	AUS	6	117	-	-	5	34	ITA	3	49	-	-	2
16	CAN	6	112	-	2	1	35	KWT	5	48	-	-	-
17	POL	6	93	-	-	3	36	ISL	4	37	-	-	-
18	MAR	6	90	-	1	2	37	LUX	2	22	-	-	-
19	TUN	6	85	-	-	1							

28th IMO 1987 - Cuba

	Country	P	TS	G	S	B		Country	P	TS	G	S	B
1	ROU	6	250	5	1	-	22	ESP	6	91	-	-	3
2	GER	6	248	4	2	-	23	MAR	6	88	-	-	3
3	USS	6	235	3	3	-	24	CUB	6	83	-	-	2
4	GDR	6	231	2	3	1	25	BEL	6	74	-	-	1
5	USA	6	220	2	3	1	26	FIN	6	70	-	-	2
6	HUN	6	218	-	5	1		IRN	6	70	-	-	1
7	BGR	6	210	1	3	2	28	NOR	6	69	-	-	-
8	CHN	6	200	2	2	2	29	COL	6	68	-	-	1
9	CZS	6	192	-	4	2	30	MNG	6	67	-	-	-
10	UNK	6	182	1	2	2	31	POL	3	55	-	-	2
11	VNM	6	172	-	1	5	32	ISL	4	45	-	-	-
12	FRA	6	154	-	3	2	33	CYP	6	42	-	-	-
13	AUT	6	150	-	2	3	34	PER	6	41	-	-	-
14	NLD	6	146	-	1	4	35	ITA	4	35	-	-	1
15	AUS	6	143	-	3	-	36	ALG	6	29	-	-	-
16	CAN	6	139	1	1	1	37	KWT	6	28	-	-	-
17	SWE	6	134	-	2	2	38	LUX	1	27	-	-	1
18	YUG	6	132	-	1	3		URY	4	27	-	-	-
19	BRA	6	116	1	-	2	40	MEX	5	17	-	-	-
20	HEL	6	111	-	-	4	41	NIC	6	13	-	-	-
21	TUR	6	94	-	-	2	42	PAN	6	7	-	-	-

29th IMO 1988 - Australia

	Country	P	TS	G	S	B	H		Country	P	TS	G	S	B	H
1	USS	6	217	4	2	-	-	26	COL	6	66	-	-	3	-
2	CHN	6	201	2	4	-	-	27	FIN	6	65	-	-	2	1
	ROU	6	201	2	4	-	-		HEL	6	65	-	-	1	3
4	GER	6	174	1	4	1	-		TUR	6	65	-	-	3	-
5	VNM	6	166	1	4	-	-	30	LUX	3	64	-	1	2	-
6	USA	6	153	-	5	1	-	31	MAR	6	62	-	-	2	1
7	GDR	5	145	1	4	-	-	32	PER	6	55	-	-	1	3
8	BGR	6	144	-	4	2	-	33	POL	3	54	-	1	-	2
9	FRA	6	128	1	1	3	1	34	NZL	6	47	-	1	-	-
10	CAN	6	124	1	1	2	1	35	ITA	4	44	-	-	1	1
11	UNK	6	121	-	3	2	-	36	ALG	5	42	-	1	-	1
12	CZS	6	120	-	2	2	1	37	MEX	6	40	-	-	1	2
13	ISR	6	115	1	-	4	1	38	BRA	6	39	-	-	-	2
	SWE	6	115	1	-	4	1	39	ISL	4	37	-	-	1	-
15	AUT	6	110	1	1	1	1	40	CUB	6	35	-	-	-	1
16	HUN	6	109	-	2	2	1	41	ESP	6	34	-	-	-	1
17	AUS	6	100	1	-	1	1	42	NOR	6	33	-	-	-	-
18	SGP	6	96	-	2	2	-	43	IRL	6	30	-	-	-	-
19	YUG	6	92	-	-	4	1	44	PHI	5	29	-	-	-	1
20	IRN	6	86	-	1	3	-	45	ARG	3	23	-	-	-	1
21	NLD	6	85	-	-	3	1		KWT	6	23	-	-	-	-
22	KOR	6	79	-	-	3	-	47	CYP	6	21	-	-	-	1
23	BEL	6	76	-	-	3	1	48	IDN	3	6	-	-	-	-
24	HKG	6	68	-	-	2	1	49	ECU	1	1	-	-	-	-
25	TUN	4	67	-	-	3	-								

30th IMO 1989 - Germany

	Country	P	TS	G	S	B	H		Country	P	TS	G	S	B	H
1	CHN	6	237	4	2	-	-	26	ISR	6	105	-	2	1	-
2	ROU	6	223	2	4	-	-	27	BEL	6	104	-	-	3	2
3	USS	6	217	3	2	1	-	28	KOR	6	97	-	1	-	4
4	GDR	6	216	3	2	1	-	29	NLD	6	92	-	1	1	2
5	USA	6	207	1	4	1	-	30	TUN	6	81	-	1	-	2
6	CZS	6	202	2	1	3	-	31	MEX	6	79	-	-	1	2
7	BGR	6	195	1	3	2	-	32	SWE	6	73	-	-	2	1
8	GER	6	187	1	3	2	-	33	CUB	6	69	-	-	1	3
9	VNM	6	183	2	1	3	-		NZL	6	69	-	-	2	2
10	HUN	6	175	-	4	1	1	35	LUX	3	65	-	1	1	-
11	YUG	6	170	1	3	1	1	36	BRA	6	64	-	-	3	-
12	POL	6	157	-	3	3	-		NOR	4	64	-	-	1	2
13	FRA	6	156	-	1	5	-	38	MAR	6	63	-	-	1	3
14	IRN	6	147	-	2	3	1	39	ESP	6	61	-	-	1	4
15	SGP	6	143	-	-	4	2	40	FIN	6	58	-	-	-	3
16	TUR	6	133	-	1	4	1	41	THA	6	54	-	-	1	2
17	HKG	6	127	-	2	1	1	42	PER	6	51	-	-	-	3
18	ITA	6	124	-	1	2	3	43	PHI	6	45	-	1	-	-
19	CAN	6	123	-	1	3	2	44	POR	6	39	-	-	-	4
20	HEL	6	122	-	1	3	2	45	IRL	6	37	-	-	-	1
	UNK	6	122	-	2	1	2	46	ISL	4	33	-	-	-	2
22	AUS	6	119	-	2	2	-	47	KWT	6	31	-	-	-	-
	COL	6	119	-	1	2	3	48	CYP	6	24	-	-	-	1
24	AUT	6	111	-	2	1	1	49	IDN	6	21	-	-	-	-
25	IND	6	107	-	-	4	1	50	VEN	4	6	-	-	-	-

31st IMO 1990 - People's Republic of China

	Country	P	TS	G	S	B	H		Country	P	TS	G	S	B	H
1	CHN	6	230	5	1	-	-	28	SWE	6	91	-	1	2	-
2	USS	6	193	3	2	1	-	29	NLD	6	90	-	1	2	2
3	USA	6	174	2	3	-	-	30	COL	6	88	-	1	2	-
4	ROU	6	171	2	2	2	-	31	NZL	6	83	-	-	2	2
5	FRA	6	168	3	1	-	1	32	KOR	6	79	-	1	1	1
6	HUN	6	162	1	3	2	-	33	THA	6	75	-	-	2	2
7	GDR	6	158	-	4	2	-		TUR	6	75	-	-	1	2
8	CZS	6	153	-	5	1	-	35	ESP	6	72	-	-	-	-
9	BGR	6	152	1	4	1	-	36	MAR	5	71	-	1	-	-
10	UNK	6	141	2	-	2	1	37	MEX	6	69	-	-	1	2
11	CAN	6	139	-	3	1	2	38	ARG	6	67	-	-	1	2
12	GER	6	138	-	2	4	-		CUB	6	67	-	-	1	1
13	ITA	6	131	1	1	4	-	40	BAH	6	65	-	-	-	1
14	IRN	6	122	-	4	-	-		IRL	6	65	-	-	1	-
15	AUS	6	121	-	2	4	-	42	HEL	6	62	-	-	1	1
	AUT	6	121	-	1	4	1	43	FIN	6	59	-	-	1	1
17	IND	6	116	1	1	2	-	44	LUX	2	58	1	-	1	-
18	NOR	6	112	-	3	1	-	45	TUN	4	55	-	1	1	-
19	PRK	6	109	-	1	3	-	46	MNG	6	54	-	-	-	3
20	JPN	6	107	-	2	1	-	47	KWT	4	53	-	-	1	1
21	POL	6	106	-	2	1	2	48	CYP	4	46	-	-	1	-
22	HKG	6	105	-	-	4	1		PHI	6	46	-	-	1	-
23	VNM	6	104	-	1	3	-	50	POR	6	44	-	-	-	-
24	BRA	6	102	1	-	2	-	51	IDN	6	40	-	-	-	-
25	YUG	6	98	-	1	2	1	52	MAC	6	32	-	-	-	-
26	ISR	6	95	-	1	3	-	53	ISL	3	30	-	-	1	-
27	SGP	6	93	-	-	2	2	54	ALG	4	29	-	-	-	-

32nd IMO 1991 - Sweden

	Country	P	TS	G	S	B	H		Country	P	TS	G	S	B	H
1	USS	6	241	4	2	-	-	29	HKG	6	91	-	-	2	1
2	CHN	6	231	4	2	-	-		NZL	6	91	-	-	2	1
3	ROU	6	225	3	2	1	-	31	MAR	6	85	-	-	1	4
4	GER	6	222	1	5	-	-		NOR	6	85	-	-	3	-
5	USA	6	212	1	4	1	-	33	HEL	6	81	-	-	2	1
6	HUN	6	209	2	3	1	-	34	CUB	6	80	-	-	2	2
7	BGR	6	192	-	3	3	-	35	MEX	6	76	-	-	1	3
8	IRN	6	191	2	1	2	-	36	ITA	6	74	-	-	1	1
	VNM	6	191	-	4	2	-	37	BRA	6	73	-	-	1	1
10	IND	6	187	-	3	3	-		NLD	6	73	-	-	1	3
11	CZS	6	186	-	4	1	1	39	TUN	4	69	-	-	2	1
12	JPN	6	180	-	3	3	-	40	ESP	6	66	-	-	1	-
13	FRA	6	175	1	1	4	-		FIN	6	66	-	-	1	1
14	CAN	6	164	1	2	2	1	42	PHI	4	64	-	-	2	1
15	POL	6	161	-	2	4	-	43	DEN	5	49	-	-	-	2
16	YUG	6	160	-	2	3	1	44	IRL	6	47	-	-	-	-
17	KOR	6	151	-	1	4	1	45	TTO	4	46	-	-	-	3
18	AUT	6	142	-	2	3	-	46	POR	6	42	-	-	-	-
	UNK	6	142	1	-	2	3	47	MNG	6	33	-	-	-	3
20	AUS	6	129	-	-	3	2	48	IDN	6	30	-	-	-	1
21	SWE	6	125	-	2	1	-		LUX	2	30	-	-	1	-
22	BEL	6	121	-	-	3	3	50	ISL	6	29	-	-	1	-
23	ISR	6	115	-	1	2	1		SUI	1	29	-	-	1	-
24	TUR	6	111	-	-	4	1	52	CYP	4	25	-	-	-	-
25	THA	6	103	-	1	1	3	53	ALG	6	20	-	-	-	1
26	COL	6	96	-	-	2	2	54	MAC	6	18	-	-	-	-
27	ARG	6	94	-	-	3	1	55	BAH	6	4	-	-	-	-
	SGP	6	94	-	1	1	1	56	PRK	6	-	disqualified			

33rd IMO 1992 - Russian Federation

	Country	P	TS	G	S	B	H		Country	P	TS	G	S	B	H
1	CHN	6	240	6	-	-	-	29	NOR	6	77	-	1	2	1
2	USA	6	181	3	3	-	-	30	NLD	6	71	-	1	-	2
3	ROU	6	177	2	2	2	-	31	AUT	6	70	-	-	3	-
4	CIS	6	176	2	3	-	1	32	ARG	6	67	-	1	1	1
5	UNK	6	168	2	2	2	-	33	TUN	4	64	1	-	1	-
6	RUS	6	158	2	2	2	-	34	TUR	6	63	-	-	2	-
7	GER	6	149	-	4	2	-	35	COL	6	55	-	-	1	-
8	HUN	6	142	1	3	1	-	36	MNG	6	51	-	-	-	2
	JPN	6	142	1	3	1	1	37	ESP	6	50	-	-	1	1
10	FRA	6	139	1	3	1	-		THA	6	50	-	1	-	-
	VNM	6	139	1	2	3	-	39	BRA	6	48	-	-	1	1
12	YUG	6	136	-	2	4	-	40	MAR	6	45	-	-	-	2
13	CZS	6	134	-	2	3	-	41	DEN	5	42	-	-	-	-
14	IRN	6	133	-	3	2	1		IRL	6	42	-	-	-	1
15	BGR	6	127	1	1	3	-	43	NZL	6	41	-	-	1	-
16	PRK	6	126	-	3	2	-	44	PHI	4	40	-	-	1	1
17	TWN	6	124	-	3	2	-	45	HEL	6	37	-	-	-	2
18	KOR	6	122	1	-	4	-	46	MAC	6	35	-	-	-	-
19	AUS	6	118	1	1	2	1		POR	6	35	-	-	1	-
20	ISR	6	108	-	2	2	1	48	CYP	6	34	-	-	1	-
21	IND	6	107	-	1	4	1	49	FIN	6	33	-	-	-	-
22	CAN	6	105	1	-	3	1	50	MEX	6	32	-	-	-	2
23	BEL	6	100	-	1	2	1	51	SUI	3	30	-	-	-	1
24	POL	6	90	-	1	3	1	52	TTO	6	26	-	-	-	-
	SWE	6	90	-	2	-	-	53	IDN	6	22	-	-	-	-
26	HKG	6	89	-	1	2	2	54	SAF	6	21	-	-	-	-
	SGP	6	89	-	1	3	-	55	CUB	3	17	-	-	-	1
28	ITA	6	83	-	-	3	1	56	ISL	3	16	-	-	-	1

34th IMO 1993 - Turkey

	Country	P	TS	G	S	B	H		Country	P	TS	G	S	B	H
1	CHN	6	215	6	-	-	-	38	BLR	4	54	-	1	1	-
2	GER	6	189	4	2	-	-	39	SWE	6	51	-	1	1	-
3	BGR	6	178	2	4	-	-	40	MAR	6	49	-	-	1	2
4	RUS	6	177	4	1	1	-	41	THA	6	47	-	-	2	-
5	TWN	6	162	1	4	1	-	42	ARG	6	46	-	1	1	-
6	IRN	6	153	2	3	1	-		SUI	4	46	-	1	1	-
7	USA	6	151	2	2	2	-	44	NOR	5	44	-	-	2	-
8	HUN	6	143	3	1	2	-	45	ESP	6	43	-	1	1	-
9	VNM	6	138	1	4	1	-		NZL	6	43	-	-	2	-
10	CZE	6	132	1	2	3	-		SVN	5	43	-	-	2	1
11	ROU	6	128	1	2	3	-	48	MKD	4	42	-	-	3	-
12	SVK	6	126	1	3	1	-	49	LTU	6	41	-	-	-	-
13	AUS	6	125	1	2	3	-	50	IRL	6	39	-	-	1	-
14	UNK	6	118	-	3	3	-	51	POR	6	35	-	-	1	-
15	IND	6	116	-	4	1	-	52	AZE	5	33	-	-	1	-
	KOR	6	116	-	3	3	-		FIN	6	33	-	-	-	-
17	FRA	6	115	2	1	1	-		PHI	6	33	-	-	1	-
18	CAN	6	113	1	1	3	-	55	HRV	6	32	-	-	1	-
	ISR	6	113	1	2	2	-	56	EST	6	31	-	-	1	-
20	JPN	6	98	-	2	3	-	57	SAF	6	30	-	-	-	-
21	UKR	6	96	-	2	3	-		TTO	6	30	-	-	-	-
22	AUT	6	87	-	1	4	-	59	MDA	6	29	-	-	-	-
23	ITA	6	86	1	-	2	-	60	KGZ	5	28	-	-	-	-
24	TUR	6	81	-	1	2	-	61	MAC	6	24	-	-	-	-
25	KAZ	6	80	-	1	3	-		MEX	6	24	-	-	1	-
26	COL	6	79	-	-	4	-		MNG	6	24	-	-	1	-
	GEO	6	79	-	1	3	1	64	ISL	4	23	-	-	-	-
28	ARM	6	78	1	1	-	-	65	LUX	1	20	-	1	-	-
	POL	6	78	-	2	1	1	66	ALB	6	18	-	-	-	-
30	SGP	6	75	-	1	3	-	67	NCY	6	17	-	-	-	-
31	LVA	6	73	-	2	1	1	68	BAH	6	16	-	-	-	-
32	DEN	6	72	-	1	3	-		KWT	6	16	-	-	-	-
33	HKG	6	70	-	-	4	1	70	IDN	6	15	-	-	-	-
34	BRA	6	60	-	-	1	2	71	BIH	2	14	-	-	1	-
35	NLD	6	58	-	-	1	1	72	ALG	5	9	-	-	-	-
36	CUB	6	56	-	1	1	1		TKM	3	9	-	-	-	-
37	BEL	6	55	-	-	1	1								

35th IMO 1994 - Hong Kong

	Country	P	TS	G	S	B	H		Country	P	TS	G	S	B	H
1	USA	6	252	6	-	-	-	36	ITA	6	102	-	-	2	3
2	CHN	6	229	3	3	-	-	37	NLD	6	99	-	-	2	3
3	RUS	6	224	3	2	1	-	38	LVA	6	98	-	-	3	1
4	BGR	6	223	3	2	1	-	39	BRA	5	95	-	2	-	3
5	HUN	6	221	1	5	-	-		GEO	6	95	-	-	2	3
6	VNM	6	207	1	5	-	-	41	SWE	6	92	-	-	1	3
7	UNK	6	206	2	2	2	-	42	HEL	6	91	-	-	1	5
8	IRN	6	203	2	2	2	-	43	HRV	6	90	-	-	2	2
9	ROU	6	198	-	5	1	-	44	EST	5	82	-	-	1	3
10	JPN	6	180	1	2	3	-	45	NOR	6	80	-	1	1	-
11	GER	6	175	1	2	3	-	46	MAC	6	75	-	1	-	3
12	AUS	6	173	-	2	3	1	47	LTU	6	73	-	-	1	1
13	KOR	6	170	-	2	4	-	48	FIN	6	70	-	-	-	4
	POL	6	170	2	-	3	1	49	IRL	6	68	-	-	-	3
	TWN	6	170	-	4	1	1	50	MKD	4	67	-	-	1	2
16	IND	6	168	-	3	3	-	51	MNG	6	65	-	1	-	3
17	UKR	6	163	1	1	2	2	52	TTO	6	63	-	-	-	1
18	HKG	6	162	-	2	4	-	53	PHI	6	53	-	-	-	1
19	FRA	6	161	1	1	3	-	54	CHI	2	52	-	1	-	1
20	ARG	6	159	-	3	1	-		MDA	6	52	-	-	1	1
21	CZE	6	154	-	2	2	2		POR	6	52	-	-	-	-
22	SVK	6	150	1	1	2	1	57	DEN	4	51	-	-	2	-
23	BLR	6	144	-	1	4	1	58	CYP	6	48	-	-	-	1
24	CAN	6	143	1	-	3	1	59	SVN	5	47	-	-	-	3
	ISR	6	143	-	1	4	1	60	IDN	6	46	-	-	-	3
26	COL	6	136	-	2	2	2	61	BIH	5	44	-	-	1	1
27	SAF	6	120	-	-	3	1	62	ESP	6	41	-	-	-	2
28	TUR	6	118	-	-	4	2	63	SUI	3	35	-	-	1	-
29	NZL	6	116	-	-	4	1	64	LUX	1	32	-	1	-	-
	SGP	6	116	-	2	-	3	65	ISL	4	29	-	-	-	1
31	AUT	6	114	1	-	-	2		MEX	6	29	-	-	-	1
32	ARM	5	110	-	-	4	1	67	KGZ	6	24	-	-	-	1
33	THA	6	106	-	-	3	1	68	CUB	1	12	-	-	-	1
34	BEL	6	105	-	-	2	4		KWT	5	12	-	-	-	-
	MAR	6	105	-	-	2	4								

36th IMO 1995 - Canada

	Country	P	TS	G	S	B	H		Country	P	TS	G	S	B	H
1	CHN	6	236	4	2	-	-	38	COL	6	100	-	1	2	2
2	ROU	6	230	4	2	-	-	39	LVA	6	97	-	1	1	2
3	RUS	6	227	4	2	-	-		SUI	5	97	-	2	-	2
4	VNM	6	220	2	4	-	-	41	SAF	6	95	-	-	2	4
5	HUN	6	210	3	1	2	-	42	MNG	6	91	-	-	1	5
6	BGR	6	207	1	4	1	-	43	AUT	6	88	-	-	1	3
7	KOR	6	203	2	3	1	-	44	BRA	6	86	1	-	-	3
8	IRN	6	202	2	3	1	-	45	NLD	6	85	-	-	2	1
9	JPN	6	183	1	3	2	-	46	NZL	6	84	-	1	1	1
10	UNK	6	180	2	1	3	-	47	BEL	6	83	-	-	1	4
11	USA	6	178	-	3	3	-	48	GEO	6	79	-	1	-	2
12	TWN	6	176	-	4	1	1	49	DEN	6	77	-	-	1	4
13	ISR	6	171	1	2	2	1	50	LTU	6	74	-	-	-	4
14	IND	6	165	-	3	3	-	51	ESP	6	72	-	-	1	3
15	GER	6	162	1	3	1	1	52	NOR	6	70	-	-	1	2
16	POL	6	161	-	1	5	-	53	IDN	6	68	-	-	1	3
17	CZE	6	154	-	1	5	-	54	HEL	6	66	-	-	1	2
	YUG	6	154	-	2	3	1	55	CUB	4	59	-	-	-	2
19	CAN	6	153	-	2	3	1	56	EST	6	55	-	-	-	2
20	HKG	6	151	-	2	3	1	57	KAZ	6	54	-	-	-	3
21	AUS	6	145	-	1	4	1	58	CYP	6	43	-	-	-	3
	SVK	6	145	-	2	2	2		MEX	6	43	-	-	1	-
23	UKR	6	140	1	1	1	2	60	SVN	5	42	-	-	-	3
24	MAR	6	138	-	1	4	1	61	IRL	6	41	-	-	-	2
25	TUR	6	134	-	2	3	-	62	MAC	6	33	-	-	-	-
26	BLR	6	131	-	1	3	2	63	TTO	6	32	-	-	-	2
	ITA	6	131	-	-	5	1	64	AZE	3	30	-	-	-	1
	SGP	6	131	-	2	2	1	65	KGZ	6	28	-	-	-	1
29	ARG	6	129	-	2	2	1		PHI	6	28	-	-	1	-
30	FRA	6	119	1	-	2	3	67	POR	6	26	-	-	-	-
31	MKD	6	117	-	1	3	-	68	ISL	4	19	-	-	-	-
32	ARM	6	111	-	2	1	1	69	BIH	6	18	-	-	-	-
	HRV	6	111	-	-	3	2	70	CHI	2	14	-	-	-	1
34	THA	6	107	-	1	2	1	71	LKA	1	10	-	-	-	1
35	SWE	6	106	-	-	2	3	72	MAS	2	1	-	-	-	-
36	FIN	6	101	-	-	3	2	73	KWT	2	-	-	-	-	-
	MDA	6	101	-	1	1	2								

37th IMO 1996 - India

	Country	P	TS	G	S	B	H		Country	P	TS	G	S	B	H
1	ROU	6	187	4	2	-	-	39	FIN	6	58	-	-	2	2
2	USA	6	185	4	2	-	-	40	SWE	6	57	-	1	1	1
3	HUN	6	167	3	2	1	-	41	MDA	5	55	-	-	2	1
4	RUS	6	162	2	3	1	-	42	AUT	6	54	-	1	-	-
5	UNK	6	161	2	4	-	-	43	SAF	6	50	-	-	2	-
6	CHN	6	160	3	2	1	-	44	MNG	6	49	-	-	2	1
7	VNM	6	155	3	1	1	-		SVN	6	49	-	-	2	-
8	KOR	6	151	2	3	-	-	46	COL	6	48	-	1	-	-
9	IRN	6	143	1	4	1	-	47	THA	6	47	-	-	1	2
10	GER	6	137	3	1	1	-	48	DEN	6	44	-	-	2	-
11	BGR	6	136	1	4	1	-		ESP	6	44	-	-	-	1
	JPN	6	136	1	3	1	-		MAC	6	44	-	-	1	1
13	POL	6	122	-	3	3	-		MKD	6	44	-	-	2	-
14	IND	6	118	1	3	1	-	52	BRA	6	36	-	-	-	1
15	ISR	6	114	1	2	2	-	53	LKA	6	34	-	-	1	-
16	CAN	6	111	-	3	3	-		MEX	6	34	-	-	-	-
17	SVK	6	108	-	2	4	-	55	EST	6	33	-	-	-	-
18	UKR	6	105	1	-	5	-	56	ISL	6	31	-	-	1	-
19	TUR	6	104	-	2	3	-	57	BIH	4	30	-	-	1	-
20	TWN	6	100	-	2	3	1	58	AZE	6	27	-	-	-	1
21	BLR	6	99	1	1	2	1	59	NLD	6	26	-	-	-	-
22	HEL	6	95	-	1	5	-	60	TTO	6	25	-	-	-	-
23	AUS	6	93	-	2	3	-	61	IRL	6	24	-	-	-	-
24	YUG	6	87	-	1	2	1	62	SUI	4	23	-	-	1	-
25	ITA	6	86	-	2	2	-	63	POR	6	21	-	-	-	1
	SGP	6	86	1	-	3	-	64	KAZ	6	20	-	-	-	-
27	HKG	6	84	-	1	4	-	65	MAR	6	19	-	-	1	-
28	CZE	6	83	-	2	1	-	66	CUB	1	16	-	-	1	-
29	ARG	6	80	-	1	3	-	67	ALB	4	15	-	-	-	-
30	GEO	6	78	1	-	2	1		KGZ	6	15	-	-	-	-
31	BEL	6	75	-	-	4	-	69	CYP	5	14	-	-	-	-
32	LTU	6	68	-	1	2	1	70	IDN	6	11	-	-	-	-
33	LVA	6	66	-	-	3	1	71	CHI	2	10	-	-	-	-
34	ARM	6	63	-	-	1	3	72	MAS	4	9	-	-	-	1
	HRV	6	63	-	1	1	-		TKM	4	9	-	-	-	-
36	FRA	6	61	-	2	-	-	74	PHI	6	8	-	-	-	-
37	NOR	6	60	-	-	3	-	75	KWT	3	1	-	-	-	-
	NZL	6	60	-	-	3	-								

38th IMO 1997 - Argentina

	Country	P	TS	G	S	B	H		Country	P	TS	G	S	B	H
1	CHN	6	223	6	-	-	-		SGP	6	88	-	-	4	1
2	HUN	6	219	4	2	-	-	43	AUT	6	86	1	-	1	2
3	IRN	6	217	4	2	-	-	44	NOR	6	79	-	-	3	-
4	RUS	6	202	3	2	1	-	45	HEL	6	75	-	1	-	3
	USA	6	202	2	4	-	-	46	KAZ	6	73	-	-	1	4
6	UKR	6	195	3	3	-	-		MKD	6	73	-	-	3	1
7	BGR	6	191	2	3	1	-	48	ITA	6	71	-	-	1	3
	ROU	6	191	2	3	1	-		NZL	6	71	-	-	2	1
9	AUS	6	187	2	3	1	-	50	SVN	6	70	-	-	2	2
10	VNM	6	183	1	5	-	-	51	LTU	6	67	-	1	1	1
11	KOR	6	164	1	4	1	-	52	THA	6	66	-	-	1	4
12	JPN	6	163	1	3	1	1	53	EST	6	64	-	-	2	1
13	GER	6	161	1	3	2	-		PER	6	64	-	-	2	3
14	TWN	6	148	-	4	2	-	55	AZE	6	56	-	-	1	1
15	IND	6	146	-	3	3	-	56	MAC	6	55	-	-	-	5
16	UNK	6	144	1	2	2	-	57	DEN	6	53	-	-	1	2
17	BLR	6	140	-	2	4	-		MDA	3	53	-	-	2	1
18	CZE	6	139	1	2	2	-		SUI	5	53	-	-	2	-
19	SWE	6	128	1	-	3	-	60	ISL	6	48	-	1	-	-
20	POL	6	125	-	2	2	2		MAR	6	48	-	-	-	2
	YUG	6	125	-	2	3	-	62	BIH	5	45	-	-	1	3
22	ISR	6	124	-	1	5	-	63	IDN	6	44	-	-	-	3
	LVA	6	124	-	1	4	1	64	ESP	6	39	-	-	-	-
24	HRV	6	121	-	1	4	1	65	TTO	6	30	-	-	-	1
25	TUR	6	119	-	1	4	-	66	CHI	6	28	-	-	-	2
26	BRA	6	117	-	1	4	1	67	UZB	3	23	-	-	-	2
27	COL	6	112	-	-	6	-	68	IRL	6	21	-	-	-	-
28	GEO	6	109	-	1	3	2	69	MAS	6	19	-	-	-	1
29	CAN	6	107	-	2	2	1		URY	6	19	-	-	-	1
30	HKG	6	106	-	-	5	-	71	ALB	3	15	-	-	-	2
	MNG	6	106	1	-	3	1		POR	5	15	-	-	-	-
32	FRA	6	105	1	-	1	2	73	PHI	2	14	-	-	-	1
	MEX	6	105	-	1	3	1	74	BOL	3	13	-	-	-	1
34	ARM	6	97	-	-	3	1	75	KGZ	3	11	-	-	-	-
	FIN	6	97	-	-	4	1	76	KWT	4	8	-	-	-	-
36	SVK	6	96	-	1	2	1		PAR	6	8	-	-	-	-
37	ARG	6	94	-	-	3	2		PRI	6	8	-	-	-	-
	NLD	6	94	-	2	-	2	79	GTM	6	7	-	-	-	-
39	SAF	6	93	1	-	2	1	80	CYP	3	5	-	-	-	-
40	CUB	6	91	-	1	2	3	81	VEN	3	4	-	-	-	-
41	BEL	6	88	-	-	3	-	82	ALG	4	3	-	-	-	-

39th IMO 1998 - Taiwan

	Country	P	TS	G	S	B	H		Country	P	TS	G	S	B	H
1	IRN	6	211	5	1	-	-	39	BEL	6	71	-	1	1	-
2	BGR	6	195	3	3	-	-	40	MKD	6	69	-	-	1	1
3	HUN	6	186	4	2	-	-	41	COL	6	66	1	-	-	2
	USA	6	186	3	3	-	-	42	THA	6	65	-	-	2	1
5	TWN	6	184	3	2	1	-	43	EST	6	63	-	1	1	-
6	RUS	6	175	2	3	1	-	44	MEX	6	62	-	1	-	1
7	IND	6	174	3	3	-	-		NLD	6	62	-	1	-	-
8	UKR	6	166	1	3	2	-	46	PER	3	60	-	2	-	1
9	VNM	6	158	1	3	2	-	47	SWE	6	58	-	-	2	-
10	YUG	6	156	-	5	-	1	48	AUT	6	57	-	-	2	1
11	ROU	6	155	3	-	2	1	49	NZL	6	50	-	-	2	-
12	KOR	6	154	2	2	2	-	50	MDA	2	45	-	1	1	-
13	AUS	6	146	-	4	2	-	51	SVN	6	44	-	-	1	2
14	JPN	6	139	1	1	3	1	52	ISL	6	42	-	-	-	3
15	CZE	6	135	-	3	3	-		MAR	6	42	-	-	-	3
16	GER	6	129	-	3	2	-	54	AZE	5	41	-	-	1	1
17	TUR	6	122	-	2	4	-	55	LTU	6	40	-	-	1	1
	UNK	6	122	-	1	4	1	56	CYP	4	39	-	-	1	2
19	BLR	6	118	-	1	4	-	57	SUI	6	37	-	-	-	2
20	CAN	6	113	1	1	2	1	58	ESP	6	36	-	-	1	1
21	POL	6	112	1	1	1	3		IRL	6	36	-	-	1	-
22	HRV	6	110	-	-	5	-		TTO	6	36	-	-	1	-
	SGP	6	110	-	1	3	2	61	NOR	6	33	-	-	-	1
24	ISR	6	104	-	-	5	-	62	MAS	6	32	-	-	-	-
25	HKG	6	102	-	1	3	1	63	FIN	6	30	-	-	-	1
26	ARM	6	100	-	2	2	-	64	MAC	5	29	-	-	-	2
	FRA	6	100	1	-	2	2	65	LUX	2	25	-	-	1	1
28	SAF	6	98	-	1	2	3	66	DEN	6	21	-	-	-	-
29	ARG	6	97	1	-	3	-	67	CUB	1	19	-	-	1	-
30	BRA	6	91	1	-	1	2	68	IDN	5	16	-	-	-	-
	MNG	6	91	-	2	2	-	69	KGZ	5	14	-	-	-	-
32	HEL	6	90	-	2	1	1	70	PHI	4	11	-	-	-	-
33	BIH	6	88	-	1	2	3		URY	6	11	-	-	-	-
	SVK	6	88	-	1	4	-	72	PAR	5	6	-	-	-	-
35	KAZ	6	81	-	-	2	3		POR	6	6	-	-	-	-
36	GEO	6	78	-	-	3	2	74	LKA	1	5	-	-	-	-
37	LVA	6	74	-	1	3	-	75	VEN	2	1	-	-	-	-
38	ITA	6	72	-	-	3	2	76	KWT	3	-	-	-	-	-

40th IMO 1999 - Romania

	Country	P	TS	G	S	B	H		Country	P	TS	G	S	B	H
1	CHN	6	182	4	2	-	-		SWE	6	66	-	-	3	-
	RUS	6	182	4	2	-	-	43	BIH	6	65	-	-	3	-
3	VNM	6	177	3	3	-	-		FIN	6	65	-	1	-	1
4	ROU	6	173	3	3	-	-	45	ARG	6	63	-	-	3	-
5	BGR	6	170	3	3	-	-	46	ESP	6	60	-	-	1	1
6	BLR	6	167	3	3	-	-	47	HEL	6	57	-	2	-	-
7	KOR	6	164	3	3	-	-		THA	6	57	-	-	3	-
8	IRN	6	159	2	4	-	-	49	COL	6	55	-	1	1	-
9	TWN	6	153	1	5	-	-		CZE	6	55	-	-	1	1
10	USA	6	150	2	3	1	-	51	LTU	6	54	-	-	2	-
11	HUN	6	147	1	4	1	-	52	MEX	6	53	-	-	1	-
12	UKR	6	136	2	2	1	-		NZL	6	53	-	-	1	-
13	JPN	6	135	2	4	-	-	54	BEL	6	51	-	-	2	-
14	YUG	6	130	1	2	3	-		DEN	5	51	-	-	2	-
15	AUS	6	116	1	1	3	1	56	MDA	6	50	-	-	1	1
16	TUR	6	109	1	1	4	-	57	MAR	6	48	-	-	1	1
17	GER	6	108	-	2	4	-	58	SVN	6	46	-	-	2	-
18	IND	6	107	-	3	3	-	59	UZB	6	42	-	-	-	-
19	POL	6	104	1	-	5	-	60	ISL	6	41	-	-	1	-
20	UNK	6	100	-	3	2	-		MAC	6	41	-	-	-	-
21	SVK	6	88	-	2	3	-	62	IRL	6	38	-	-	1	-
22	LVA	6	86	1	1	-	-	63	MAS	6	37	-	-	-	-
23	ITA	6	82	-	1	2	-	64	CYP	6	35	-	-	-	1
24	SUI	6	79	-	1	3	-		IDN	6	35	-	-	-	-
25	ISR	6	78	-	-	5	1	66	ALB	5	34	-	-	-	-
	MNG	6	78	-	2	1	-		AZE	6	34	-	-	1	-
27	CUB	6	77	-	1	4	-	68	TTO	6	33	-	-	-	-
	SAF	6	77	-	1	1	-	69	EST	4	30	-	-	1	-
29	AUT	6	75	-	1	2	-	70	POR	6	29	-	-	-	-
	BRA	6	75	-	-	4	-	71	LUX	2	26	-	-	1	-
31	CAN	6	74	-	-	3	-	72	URY	5	25	-	-	-	-
	NLD	6	74	-	-	4	-	73	PHI	4	24	-	-	-	-
33	FRA	6	73	-	1	2	1	74	TUN	4	22	-	-	-	-
	HKG	6	73	-	-	4	1	75	GTM	6	19	-	-	-	-
35	KAZ	6	72	-	-	4	1	76	KGZ	5	15	-	-	-	-
36	MKD	6	71	-	-	5	-	77	TKM	2	13	-	-	-	-
	SGP	6	71	-	-	4	-	78	KWT	4	10	-	-	-	-
38	GEO	6	68	-	1	1	1		PER	2	10	-	-	-	-
39	ARM	6	67	-	-	3	-	80	VEN	2	8	-	-	-	-
	NOR	6	67	-	1	2	-	81	LKA	1	6	-	-	-	-
41	HRV	6	66	-	-	2	-								

41st IMO 2000 - Republic of Korea

	Country	P	TS	G	S	B	H		Country	P	TS	G	S	B	H
1	CHN	6	218	6	-	-	-	42	CZE	6	65	-	1	3	-
2	RUS	6	215	5	1	-	-	43	MKD	6	63	-	1	2	-
3	USA	6	184	3	3	-	-	44	COL	6	61	-	-	2	-
4	KOR	6	172	3	3	-	-		CUB	6	61	-	-	2	2
5	BGR	6	169	2	3	1	-	46	LVA	6	60	-	-	3	-
	VNM	6	169	3	2	1	-		NLD	6	60	-	-	2	1
7	BLR	6	165	2	2	2	-	48	BRA	6	58	-	-	3	1
8	TWN	6	164	3	2	1	-		FRA	6	58	-	-	3	1
9	HUN	6	156	1	5	-	-	50	ITA	6	57	-	-	3	-
10	IRN	6	155	2	3	1	-	51	IDN	6	54	-	-	2	-
11	ISR	6	139	2	1	3	-	52	FIN	6	52	-	-	3	1
	ROU	6	139	1	3	2	-	53	BEL	6	51	-	-	2	1
13	UKR	6	135	2	2	-	1		LUX	4	51	-	-	2	1
14	IND	6	132	-	5	1	-	55	MAR	6	48	-	-	1	2
15	JPN	6	125	1	2	3	-	56	HEL	6	46	-	-	1	1
16	AUS	6	122	1	3	1	-	57	NOR	6	45	-	-	1	-
17	CAN	6	112	1	2	1	1	58	EST	6	42	-	-	1	1
18	SVK	6	111	-	2	3	-	59	TTO	6	40	-	-	-	2
	TUR	6	111	-	3	1	1	60	ISL	6	37	-	-	-	1
20	ARM	6	108	-	2	3	-	61	DEN	6	36	-	-	1	1
	GER	6	108	1	1	2	1	62	LTU	6	34	-	-	1	1
22	UNK	6	96	-	2	4	-		NZL	6	34	-	-	-	1
23	YUG	6	93	-	1	3	-	64	AZE	6	32	-	-	-	2
24	KAZ	6	91	-	1	4	1		CYP	6	32	-	-	-	1
25	ARG	6	88	-	1	4	-		MAS	3	32	-	-	2	-
26	MDA	5	84	-	2	3	-		PER	4	32	-	-	-	3
27	SAF	6	81	-	-	4	1	68	ESP	6	29	-	-	-	1
28	HKG	6	80	-	1	2	1	69	IRL	6	28	-	-	-	1
29	BIH	6	78	-	-	4	-	70	PHI	4	23	-	-	-	1
	THA	6	78	-	1	3	-		URY	3	23	-	-	-	1
31	SWE	6	77	-	2	-	2	72	LKA	3	21	-	-	-	1
32	MEX	6	75	-	1	3	1		POR	6	21	-	-	-	-
	POL	6	75	-	1	2	1	74	ECU	6	19	-	-	-	-
34	HRV	6	73	-	-	4	1	75	ALB	5	17	-	-	-	-
	SVN	6	73	-	1	1	2	76	KGZ	4	16	-	-	1	-
36	GEO	6	72	-	1	-	4		MAC	6	16	-	-	-	-
37	SGP	6	71	-	1	2	-	78	KWT	4	12	-	-	-	-
38	UZB	6	70	-	-	2	4	79	GTM	6	11	-	-	-	-
39	AUT	6	68	-	2	1	-		VEN	2	11	-	-	-	-
40	MNG	6	67	-	-	4	-	81	BRU	2	8	-	-	-	-
	SUI	4	67	-	1	2	-		PRI	6	8	-	-	-	-

42nd IMO 2001 - United States of America

	Country	P	TS	G	S	B	H		Country	P	TS	G	S	B	H
1	CHN	6	225	6	-	-	-	43	MKD	6	59	-	-	2	-
2	RUS	6	196	5	1	-	-	44	NZL	6	58	-	1	1	-
	USA	6	196	4	2	-	-	45	CZE	6	57	-	-	2	2
4	BGR	6	185	3	3	-	-	46	ITA	6	56	-	-	2	2
	KOR	6	185	3	3	-	-		MEX	6	56	-	-	2	1
6	KAZ	6	168	4	1	-	1	48	SVK	6	54	-	-	2	-
7	IND	6	148	2	2	2	-	49	VEN	5	53	-	1	1	1
8	UKR	6	143	1	5	-	-	50	NOR	6	48	-	1	1	-
9	TWN	6	141	1	5	-	-	51	BIH	6	47	-	1	1	-
10	VNM	6	139	1	4	-	1	52	MAR	6	45	-	-	1	2
11	TUR	6	136	1	3	2	-	53	ARM	5	44	-	-	2	-
12	BLR	6	135	1	2	3	-	54	NLD	6	42	-	-	2	1
13	JPN	6	134	1	3	2	-	55	AUT	6	41	-	-	1	1
14	GER	6	131	1	3	1	-	56	LTU	6	39	-	-	1	-
15	ROU	6	129	1	2	2	1	57	SUI	6	38	-	-	2	-
16	BRA	6	120	-	4	2	-	58	ESP	6	37	-	-	1	2
17	ISR	6	113	1	2	1	1	59	IDN	6	36	-	1	-	1
18	IRN	6	111	-	2	4	-		MAS	6	36	-	-	-	3
19	HKG	6	107	-	2	4	-		TTO	6	36	-	-	2	-
	POL	6	107	-	3	1	1		TUN	6	36	-	-	1	-
21	HUN	6	104	-	2	3	-	63	FIN	6	32	-	-	1	-
22	ARG	6	103	-	3	2	-		IRL	6	32	-	-	1	-
	THA	6	103	-	2	2	-		MAC	6	32	-	-	1	-
24	CAN	6	100	1	-	4	-	66	TKM	5	29	-	-	-	1
25	AUS	6	97	1	-	4	-	67	SVN	6	27	-	-	-	-
26	CUB	6	92	1	1	3	-	68	BEL	6	25	-	-	-	-
27	UZB	6	91	-	1	3	2		DEN	6	25	-	-	-	1
28	FRA	6	88	-	2	3	-		SWE	6	25	-	-	1	-
29	SGP	6	87	-	1	4	-	71	LKA	4	21	-	-	1	-
30	HEL	6	86	-	1	3	1	72	ALB	5	20	-	-	-	1
31	MNG	6	79	-	2	2	-	73	AZE	3	19	-	-	1	-
	UNK	6	79	-	1	3	1	74	ISL	6	18	-	-	-	-
	YUG	6	79	-	1	3	-	75	PHI	6	16	-	-	-	-
34	CYP	6	78	-	-	4	2	76	GTM	3	12	-	-	-	1
35	HRV	6	76	-	1	2	2	77	URY	2	8	-	-	-	-
36	SAF	6	75	-	1	3	-	78	POR	6	6	-	-	-	-
37	EST	6	72	-	1	3	-	79	KGZ	5	5	-	-	-	-
38	GEO	6	71	-	1	3	-	80	LUX	2	4	-	-	-	-
	LVA	6	71	-	1	2	2	81	KWT	4	3	-	-	-	-
40	MDA	5	70	-	2	1	1	82	PAR	5	2	-	-	-	-
41	PER	6	67	-	-	4	1	83	ECU	6	-	-	-	-	-
42	COL	6	64	-	-	4	-								

43rd IMO 2002 - United Kingdom

	Country	P	TS	G	S	B	H		Country	P	TS	G	S	B	H
1	CHN	6	212	6	-	-	-	43	MKD	6	73	-	1	1	1
2	RUS	6	204	6	-	-	-	44	NOR	6	72	1	-	1	1
3	USA	6	171	4	1	-	1	45	HRV	6	70	-	-	2	1
4	BGR	6	167	3	2	1	-	46	MEX	6	67	-	-	3	-
5	VNM	6	166	3	1	2	-	47	HEL	6	62	-	-	2	-
6	KOR	6	163	1	5	-	-	48	MDA	6	60	-	-	2	-
7	TWN	6	161	1	4	1	-		SWE	6	60	-	-	2	1
8	ROU	6	157	2	3	1	-		UZB	6	60	-	-	-	2
9	IND	6	156	1	3	2	-	51	PER	5	59	-	-	2	-
10	GER	6	144	2	1	2	1	52	BEL	6	58	-	-	1	3
11	IRN	6	143	-	4	2	-		VEN	5	58	-	1	1	1
12	CAN	6	142	1	3	1	1	54	NLD	6	55	-	-	1	1
	HUN	6	142	1	2	3	-	55	DEN	6	53	-	-	-	3
14	BLR	6	135	1	2	3	-	56	AUT	6	50	-	-	1	3
	TUR	6	135	1	1	4	-		MAC	6	50	-	-	3	-
16	JPN	6	133	1	3	1	-	58	SVN	6	46	-	-	1	1
	KAZ	6	133	-	3	3	-	59	TKM	6	45	-	-	1	1
18	ISR	6	130	-	3	3	-	60	ESP	6	44	-	-	1	1
19	FRA	6	127	-	2	3	-		SUI	6	44	-	-	1	2
20	UKR	6	124	1	3	-	-	62	BIH	6	42	-	-	1	-
21	BRA	6	123	-	1	5	-	63	MAR	6	39	-	-	1	1
	POL	6	123	-	4	1	1	64	IDN	6	38	-	-	1	1
	THA	6	123	-	2	2	2	65	AZE	6	37	-	-	1	1
24	HKG	6	120	1	2	2	-	66	ISL	6	36	-	-	-	3
25	SVK	6	119	-	2	4	-	67	ARM	6	33	-	-	-	1
26	AUS	6	117	1	2	1	1	68	CYP	6	29	-	-	-	-
27	UNK	6	116	-	2	2	-	69	MAS	6	26	-	-	-	1
28	CZE	6	115	-	2	3	-	70	ALB	6	25	-	-	1	-
29	YUG	6	114	-	1	5	-		IRL	6	25	-	-	-	1
30	SGP	6	112	-	2	2	1	72	TTO	6	22	-	-	-	-
31	ARG	6	96	-	-	5	-		TUN	6	22	-	-	-	1
32	SAF	6	90	-	1	3	-	74	PHI	5	18	-	-	-	1
33	ITA	6	88	-	-	5	1	75	KGZ	4	17	-	-	-	1
34	GEO	6	84	-	-	2	2		PRI	6	17	-	-	-	-
35	MNG	6	82	-	-	3	-	77	LKA	4	16	-	-	-	1
	NZL	6	82	1	-	-	4	78	POR	6	15	-	-	-	-
37	COL	6	81	-	-	3	3	79	LUX	2	12	-	-	-	1
38	FIN	6	79	-	-	3	3	80	PAR	2	11	-	-	-	1
39	CUB	6	78	-	-	2	3	81	GTM	3	4	-	-	-	-
40	EST	6	75	-	2	-	2	82	ECU	6	3	-	-	-	-
	LVA	6	75	-	1	2	2	83	KWT	4	2	-	-	-	-
42	LTU	6	74	-	1	2	1	84	URY	1	1	-	-	-	-

44th IMO 2003 - Japan

	Country	P	TS	G	S	B	H		Country	P	TS	G	S	B	H
1	BGR	6	227	6	-	-	-	42	NOR	6	62	-	1	-	2
2	CHN	6	211	5	1	-	-	43	ARM	6	61	-	-	3	2
3	USA	6	188	4	2	-	-		BIH	6	61	-	-	2	2
4	VNM	6	172	2	3	1	-	45	SAF	6	60	-	-	3	-
5	RUS	6	167	3	2	1	-	46	ESP	6	59	-	-	1	4
6	KOR	6	157	2	4	-	-	47	MKD	6	54	-	-	2	3
7	ROU	6	143	1	4	1	-	48	SWE	6	52	-	-	1	3
8	TUR	6	133	1	3	1	1	49	ITA	6	50	-	-	1	4
9	JPN	6	131	1	3	2	-		KGZ	6	50	-	-	2	2
10	HUN	6	128	1	3	1	1		LVA	6	50	-	-	1	2
	UNK	6	128	1	2	3	-	52	LTU	6	49	-	-	2	2
12	CAN	6	119	2	-	3	-		UZB	6	49	-	1	1	1
	KAZ	6	119	1	2	2	1	54	EST	6	47	-	-	-	3
14	UKR	6	118	1	2	3	-	55	FIN	6	43	-	-	1	2
15	IND	6	115	-	4	1	1		MAR	6	43	-	-	-	5
16	TWN	6	114	1	2	2	1		NZL	6	43	-	-	-	3
17	GER	6	112	1	2	1	2	58	MAC	6	40	-	-	2	-
	IRN	6	112	-	3	2	1	59	AUT	6	38	-	-	-	3
19	BLR	6	111	1	2	2	-	60	PER	4	37	-	-	1	2
	THA	6	111	1	1	3	1		TKM	4	37	-	-	1	3
21	ISR	5	103	-	2	3	-	62	ISL	6	33	-	-	1	1
22	POL	6	102	1	2	-	2		TTO	6	33	-	-	-	2
23	SCG	6	101	-	3	1	2	64	NLD	6	30	-	-	-	-
24	FRA	6	95	-	2	2	1	65	URY	5	29	-	-	-	2
25	MNG	6	93	-	1	3	1	66	DEN	5	27	-	-	-	2
26	AUS	6	92	-	2	2	2	67	MAS	5	26	-	-	-	1
	BRA	6	92	-	1	3	2		SUI	6	26	-	-	-	1
28	ARG	6	91	1	1	2	1	69	LUX	2	25	-	-	1	1
	HKG	6	91	-	2	2	1	70	ALB	4	23	-	-	-	1
30	HEL	6	88	-	1	4	1		CYP	6	23	-	-	-	1
	MDA	6	88	-	1	2	3		PRI	3	23	-	-	1	-
32	GEO	6	86	-	1	2	3	73	POR	6	22	-	-	-	1
33	HRV	6	80	-	-	3	3	74	IRL	6	21	-	-	-	1
34	CZE	6	79	-	1	2	3	75	SVN	6	18	-	-	-	1
35	SVK	6	77	-	-	4	2	76	CUB	1	14	-	-	1	-
36	SGP	6	71	-	-	2	3	77	ECU	6	11	-	-	-	1
37	BEL	6	70	-	1	1	3	78	VEN	3	10	-	-	-	-
	IDN	6	70	-	-	2	4	79	PHI	6	9	-	-	-	-
39	COL	6	67	-	-	3	2	80	KWT	3	8	-	-	-	-
40	AZE	6	66	-	1	1	3	81	LKA	4	4	-	-	-	-
41	MEX	6	64	-	-	3	1	82	PAR	1	-	-	-	-	-

45th IMO 2004 - Greece

	Country	P	TS	G	S	B	H		Country	P	TS	G	S	B	H
1	CHN	6	220	6	-	-	-	44	EST	6	85	-	-	2	3
2	USA	6	212	5	1	-	-	45	UZB	6	79	-	-	3	2
3	RUS	6	205	4	1	1	-	46	SWE	6	75	-	-	3	1
4	VNM	6	196	4	2	-	-	47	AZE	6	72	-	1	-	-
5	BGR	6	194	3	3	-	-	48	MKD	6	71	-	-	1	4
6	TWN	6	190	3	3	-	-	49	ITA	6	69	-	-	2	1
7	HUN	6	187	2	3	1	-		SVN	6	69	-	-	2	2
8	JPN	6	182	2	4	-	-	51	LTU	6	65	-	-	-	5
9	IRN	6	178	1	5	-	-	52	KGZ	6	63	-	-	1	1
10	ROU	6	176	1	4	1	-		LVA	6	63	-	-	1	2
11	UKR	6	174	1	5	-	-	54	IDN	6	61	-	-	1	3
12	KOR	6	166	2	2	2	-	55	ALB	6	57	-	-	1	1
13	BLR	6	154	-	4	2	-		ESP	6	57	-	-	1	3
14	IND	6	151	-	4	2	-		SUI	6	57	-	-	2	1
15	ISR	6	147	1	1	4	-	58	NZL	6	56	-	-	2	-
16	POL	6	142	2	1	1	1	59	AUT	6	55	-	-	1	1
17	MDA	6	140	2	-	4	-		NOR	6	55	-	-	-	-
18	SGP	6	139	-	3	3	-	61	NLD	6	53	-	-	-	2
19	MNG	6	135	-	3	2	-	62	TKM	6	52	-	-	2	1
20	UNK	6	134	1	1	4	-	63	CYP	6	49	-	-	1	1
21	BRA	6	132	-	2	4	-		FIN	6	49	-	-	1	-
	CAN	6	132	1	-	3	2		PER	3	49	-	-	2	-
	KAZ	6	132	2	-	2	-	66	IRL	6	48	-	-	1	-
	SCG	6	132	-	2	3	-	67	URY	6	47	-	-	-	-
25	GER	6	130	-	3	1	2	68	DEN	6	46	-	-	1	-
26	HEL	6	126	-	2	3	1	69	PRI	5	43	-	1	-	1
27	AUS	6	125	1	1	2	1	70	BIH	6	40	-	-	-	2
28	GEO	6	123	-	-	5	1	71	LUX	3	36	-	1	-	1
29	COL	6	122	-	2	2	-	72	ISL	6	35	-	-	-	1
30	HKG	6	120	-	2	2	2	73	MAS	6	34	-	-	1	1
31	SVK	6	119	-	3	-	3	74	LKA	6	33	-	-	-	2
32	TUR	6	118	-	2	3	-	75	TUN	6	31	-	-	-	1
33	SAF	6	110	-	3	1	1	76	TTO	5	29	-	-	-	1
34	CZE	6	109	-	2	2	-	77	POR	6	26	-	-	-	2
35	THA	6	99	-	-	4	2	78	CUB	1	17	-	-	1	-
36	ARM	6	98	-	-	4	1	79	PHI	5	16	-	-	-	-
37	MEX	6	96	-	-	3	1	80	VEN	2	15	-	-	-	1
38	FRA	6	94	-	-	4	1	81	ECU	6	14	-	-	-	1
39	ARG	6	92	1	-	2	1	82	MOZ	3	13	-	-	-	-
40	HRV	6	89	-	-	3	3		PAR	3	13	-	-	-	1
41	MAR	6	88	-	-	3	3	84	KWT	6	5	-	-	-	-
42	BEL	6	86	-	1	2	1	85	SAU	6	4	-	-	-	-
	MAC	6	86	-	-	2	2								

46th IMO 2005 - Mexico

	Country	P	TS	G	S	B	H
1	CHN	6	235	5	1	-	-
2	USA	6	213	4	2	-	-
3	RUS	6	212	4	2	-	-
4	IRN	6	201	2	4	-	-
5	KOR	6	200	3	3	-	-
6	ROU	6	191	4	1	1	-
7	TWN	6	190	3	2	1	-
8	JPN	6	188	3	1	2	-
9	HUN	6	181	2	3	1	-
	UKR	6	181	2	2	2	-
11	BGR	6	173	2	3	1	-
12	GER	6	163	1	3	2	-
13	UNK	6	159	1	3	2	-
14	SGP	6	145	-	4	2	-
15	VNM	6	143	-	3	3	-
16	CZE	6	139	1	2	2	-
17	HKG	6	138	1	3	1	-
18	BLR	6	136	1	3	1	-
19	CAN	6	132	1	2	2	-
20	SVK	6	131	-	4	2	-
21	MDA	6	130	1	2	2	-
	TUR	6	130	-	4	1	1
23	THA	6	128	-	4	2	-
24	ITA	6	120	-	2	4	-
25	AUS	6	117	-	-	6	-
26	KAZ	6	112	-	2	3	1
27	COL	6	105	-	2	2	-
	POL	6	105	-	1	5	-
29	PER	6	104	-	-	6	-
30	ISR	6	99	-	2	3	1
31	MEX	6	91	-	-	4	2
32	FRA	6	83	-	-	4	2
33	ARM	6	82	-	-	5	-
	BRA	6	82	1	-	1	2
	HRV	6	82	-	1	2	2
36	IND	6	81	-	1	1	3
37	GEO	6	80	-	-	4	1
38	NZL	6	77	-	1	2	2
39	SCG	6	75	-	-	3	1
40	AUT	6	74	-	-	2	3
	BEL	6	74	-	1	1	2
42	IDN	6	70	-	-	3	-
	SUI	6	70	-	1	1	3
44	DEN	6	69	-	-	4	-
45	EST	6	68	-	-	3	-
46	ARG	6	65	-	1	2	-
47	LVA	6	62	-	-	2	2
	NLD	6	62	-	-	2	2
49	AZE	6	59	-	-	2	3
50	HEL	6	58	-	-	2	1
51	IRL	6	55	-	1	-	-
52	CUB	4	54	-	-	3	1
53	LTU	6	53	-	-	1	3
54	MKD	6	50	-	-	2	-
55	BIH	6	49	-	-	2	1
	FIN	6	49	-	-	2	-
	SVN	6	49	-	1	-	2
58	ESP	6	46	-	-	1	2
	KGZ	6	46	-	-	2	1
60	ALB	6	44	-	1	-	1
61	SWE	6	42	-	-	-	2
62	SAF	6	39	-	-	-	3
63	MAC	6	38	-	-	1	2
	NOR	6	38	-	-	-	3
65	CRI	6	37	-	-	-	2
	URY	5	37	-	-	1	1
67	LKA	6	32	-	-	1	1
68	PHI	6	30	-	-	-	3
69	POR	6	27	-	-	-	1
70	SLV	6	25	-	-	-	2
71	ISL	6	23	-	-	1	-
72	MAR	6	18	-	-	-	1
	TKM	3	18	-	-	1	-
74	ECU	6	17	-	-	1	-
75	MAS	6	15	-	-	-	-
	VEN	2	15	-	-	-	1
77	CYP	6	14	-	-	-	1
78	TTO	6	13	-	-	-	-
79	PAR	6	12	-	-	-	-
80	PAK	6	11	-	-	-	-
81	TUN	3	9	-	-	-	-
82	PRI	6	8	-	-	-	-
83	GTM	3	6	-	-	-	-
84	LIE	3	4	-	-	-	-
85	BGD	6	3	-	-	-	-
	KWT	5	3	-	-	-	-
	LUX	2	3	-	-	-	-
	SAU	5	3	-	-	-	-
	TJK	3	3	-	-	-	-
90	MOZ	5	2	-	-	-	-
91	BOL	2	-	-	-	-	-

47th IMO 2006 - Slovenia

	Country	P	TS	G	S	B	H		Country	P	TS	G	S	B	H
1	CHN	6	214	6	-	-	-		MNG	6	80	-	-	2	4
2	RUS	6	174	3	3	-	-	47	POR	6	78	-	-	3	1
3	KOR	6	170	4	2	-	-	48	AZE	6	77	-	1	1	4
4	GER	6	157	4	-	2	-		CZE	6	77	-	-	3	3
5	USA	6	154	2	4	-	-	50	ALB	6	76	-	1	1	2
6	ROU	6	152	3	1	2	-		COL	6	76	-	-	2	3
7	JPN	6	146	2	3	1	-	52	BEL	6	75	-	-	1	4
8	IRN	6	145	3	3	-	-		LVA	6	75	-	-	3	2
9	MDA	6	140	2	1	3	-	54	HRV	6	72	-	1	1	2
10	TWN	6	136	1	5	-	-	55	LKA	5	71	-	-	3	2
11	POL	6	133	1	2	3	-	56	HEL	6	69	-	-	2	3
12	ITA	6	132	2	2	-	1	57	UZB	6	68	-	-	2	3
13	VNM	6	131	2	2	2	-	58	NZL	6	66	-	-	2	2
14	HKG	6	129	1	3	2	-	59	ISL	6	63	-	-	1	2
15	CAN	6	123	-	5	1	-		MAC	6	63	-	-	2	1
	THA	6	123	1	3	2	-	61	TKM	5	59	-	1	1	1
17	HUN	6	122	-	5	1	-	62	MKD	6	57	-	-	1	3
18	SVK	6	118	1	2	3	-		NLD	6	57	-	-	-	5
19	TUR	6	117	-	4	1	1		SAF	6	57	-	-	-	5
	UNK	6	117	-	4	1	1	65	MAR	6	55	-	-	-	4
21	BGR	6	116	-	4	1	1	66	NOR	6	52	-	-	1	2
22	UKR	6	114	1	2	2	1	67	IRL	6	49	-	-	-	4
23	BLR	6	111	-	3	2	1	68	PAR	4	47	-	1	-	1
24	MEX	6	110	1	2	1	1	69	DEN	6	45	-	-	-	1
25	ISR	6	109	-	3	1	2	70	ECU	6	40	-	-	1	2
26	AUS	6	108	-	3	2	1		MAS	6	40	-	-	1	1
27	SGP	6	100	-	2	3	1	72	TJK	6	35	-	-	-	3
28	FRA	6	99	1	-	3	2	73	TTO	6	34	-	-	-	2
29	BRA	6	96	-	-	6	-		VEN	4	34	-	-	-	3
30	ARG	6	95	-	2	2	1	75	PAN	4	33	-	-	-	2
	KAZ	6	95	-	1	4	1	76	PAK	5	32	-	-	-	1
	SUI	6	95	1	1	-	4	77	KGZ	6	31	-	-	-	2
33	GEO	6	94	-	1	3	2	78	CRI	2	27	-	-	1	1
	LTU	6	94	-	1	2	3		SLV	3	27	-	-	-	2
35	IND	6	92	-	-	5	1	80	BGD	4	22	-	-	-	2
36	ARM	6	90	-	1	1	4	81	CYP	6	19	-	-	-	1
	SVN	6	90	-	1	3	2	82	LUX	2	12	-	-	-	1
38	SRB	6	88	-	-	5	1		URY	2	12	-	-	-	1
39	FIN	6	86	-	-	4	2	84	NGA	6	11	-	-	-	-
40	PER	6	85	-	1	1	4		PRI	6	11	-	-	-	-
41	BIH	6	84	-	1	2	3	86	BOL	2	5	-	-	-	-
42	AUT	6	83	-	-	3	3		KWT	4	5	-	-	-	-
43	SWE	6	82	-	-	3	3	88	SAU	4	3	-	-	-	-
44	ESP	6	80	-	1	2	2	89	LIE	1	2	-	-	-	-
	EST	6	80	-	-	2	2	90	MOZ	3	-	-	-	-	-

48th IMO 2007 - Vietnam

	Country	P	TS	G	S	B	H		Country	P	TS	G	S	B	H
1	RUS	6	184	5	1	-	-	48	ARM	6	73	-	1	1	4
2	CHN	6	181	4	2	-	-		MAC	6	73	-	1	1	4
3	KOR	6	168	2	4	-	-	50	ISR	6	71	-	-	3	3
	VNM	6	168	3	3	-	-		NZL	6	71	-	-	3	2
5	USA	6	155	2	3	1	-	52	AZE	6	69	-	-	3	1
6	JPN	6	154	2	4	-	-		BIH	6	69	-	1	-	5
	UKR	6	154	3	1	2	-		IDN	6	69	-	1	-	4
8	PRK	6	151	1	4	-	1	55	MKD	6	68	-	-	3	1
9	BGR	6	149	2	3	1	-	56	NLD	6	65	-	-	1	3
	TWN	6	149	2	3	1	-	57	EST	6	64	-	-	1	4
11	ROU	6	146	1	4	1	-	58	ALB	6	59	-	-	1	5
12	HKG	6	143	-	5	1	-		SUI	6	59	-	-	1	3
	IRN	6	143	1	3	2	-	60	LVA	6	58	-	-	-	4
14	THA	6	133	1	3	2	-	61	FIN	6	55	-	1	-	2
15	GER	6	132	1	3	1	1	62	POR	6	52	-	-	1	1
16	HUN	6	129	-	5	-	1	63	IRL	6	51	-	-	1	3
17	TUR	6	124	1	2	2	1		TKM	6	51	-	-	-	5
18	POL	6	122	1	2	2	-	65	DEN	6	50	-	-	1	3
19	BLR	6	119	1	1	4	-	66	ESP	6	48	-	-	2	1
20	MDA	6	118	-	3	2	-	67	KGZ	5	43	-	-	1	3
21	ITA	6	116	1	1	3	1	68	SAF	6	42	-	-	-	4
22	AUS	6	110	-	1	4	1	69	CYP	6	41	-	-	-	2
23	SRB	6	107	1	-	4	-	70	TTO	6	39	-	-	-	4
24	BRA	6	106	-	2	3	1	71	TJK	6	37	-	-	1	2
25	IND	6	103	-	3	-	3	72	CRI	5	36	-	-	1	1
26	GEO	6	102	1	1	1	3	73	ISL	6	35	-	-	-	1
27	CAN	6	98	-	1	3	1	74	ECU	6	34	-	-	1	2
28	KAZ	6	95	-	1	3	2		LUX	3	34	-	-	1	2
	UNK	6	95	1	-	3	2		MAS	6	34	-	-	1	2
30	COL	6	93	-	1	3	1		SLV	4	34	-	-	-	3
31	LTU	6	92	1	-	2	1	78	PAK	6	32	-	-	1	1
32	PER	6	91	-	1	2	3		PAR	4	32	-	-	-	3
33	HEL	6	89	-	1	3	2	80	BGD	5	31	-	-	-	3
34	MNG	6	88	-	2	1	3	81	MAR	6	28	-	-	-	2
	UZB	6	88	-	1	3	2	82	KHM	4	26	-	-	-	3
36	SGP	6	87	-	-	5	-	83	LKA	6	25	-	-	-	1
37	MEX	6	86	-	-	4	2	84	PHI	6	21	-	-	-	1
	SVK	6	86	-	-	4	2	85	NGA	6	20	-	-	-	1
39	SVN	6	85	-	-	5	1	86	MNE	3	17	-	-	-	1
40	CZE	6	82	-	-	5	1	87	CUB	1	16	-	-	1	-
41	SWE	6	81	-	-	4	2	88	LIE	2	14	-	-	1	-
42	AUT	6	80	-	1	3	-		VEN	3	14	-	-	-	1
43	FRA	6	79	1	-	2	-	90	PRI	3	7	-	-	-	-
	NOR	6	79	-	1	1	1	91	SAU	4	5	-	-	-	-
45	BEL	6	78	-	-	3	2	92	CHI	4	4	-	-	-	-
46	HRV	6	76	-	-	2	4	93	BOL	2	2	-	-	-	-
47	ARG	6	75	-	1	1	3								

49th IMO 2008 - Spain

	Country	P	TS	G	S	B	H		Country	P	TS	G	S	B	H
1	CHN	6	217	5	1	-	-	50	BIH	6	68	-	-	3	1
2	RUS	6	199	6	-	-	-		SUI	6	68	-	1	1	2
3	USA	6	190	4	2	-	-		SVN	6	68	-	-	2	1
4	KOR	6	188	4	2	-	-	53	SWE	6	67	-	1	-	3
5	IRN	6	181	1	5	-	-	54	DEN	6	66	-	2	-	1
6	THA	6	175	2	3	1	-	55	CRI	6	65	-	-	2	3
7	PRK	6	173	2	4	-	-		MAS	6	65	-	1	-	4
8	TUR	6	170	3	1	2	-	57	AUT	6	63	-	-	1	4
9	TWN	6	168	2	4	-	-	58	NOR	6	62	1	-	-	1
10	HUN	6	165	2	3	1	-	59	BEL	6	61	-	1	1	1
11	JPN	6	163	2	3	1	-		MKD	6	61	-	-	2	-
12	VNM	6	159	2	2	2	-	61	LUX	5	60	-	-	2	2
13	POL	6	157	2	3	1	-		TJK	6	60	-	-	1	3
14	BGR	6	154	2	1	3	-	63	LVA	6	58	-	1	-	2
15	UKR	6	153	2	2	2	-		MAC	6	58	-	-	2	1
16	BRA	6	152	-	5	1	-		MAR	6	58	-	-	1	2
17	PER	6	141	1	3	2	-	66	ARM	6	56	-	-	-	4
	ROU	6	141	-	4	2	-	67	POR	6	55	-	-	2	1
19	AUS	6	140	-	5	1	-	68	ALB	6	53	-	-	1	1
20	GER	6	139	1	2	3	-	69	CHI	3	49	-	1	1	1
	SRB	6	139	1	3	-	2	70	IRL	6	45	-	-	-	2
22	CAN	6	135	-	2	4	-	71	CYP	6	42	-	-	1	1
23	UNK	6	133	-	4	2	-		NZL	6	42	-	-	-	3
24	ITA	6	132	-	3	3	-	73	EST	6	41	-	-	1	1
25	KAZ	6	128	1	2	3	-	74	FIN	6	40	-	-	1	1
26	BLR	6	125	-	3	2	-	75	BGD	4	33	-	-	-	1
27	ISR	6	120	1	1	2	2	76	ISL	5	31	-	-	1	-
28	HKG	6	107	-	3	1	1		SLV	4	31	-	-	-	3
29	MNG	6	106	-	2	1	2	78	LKA	6	29	-	-	-	1
30	FRA	6	104	-	1	4	1	79	KGZ	5	28	-	-	-	1
31	IND	6	103	-	-	5	1		TTO	6	28	-	-	1	-
32	SGP	6	98	-	1	3	1	81	CUB	1	27	-	1	-	-
33	NLD	6	94	-	2	2	-	82	ECU	6	26	-	-	-	1
	UZB	6	94	-	-	4	-	83	KHM	6	25	-	-	-	1
35	LTU	6	92	-	1	2	3	84	MNE	3	24	-	-	-	2
36	IDN	6	88	-	1	2	2		PAR	4	24	-	-	1	-
37	MEX	6	87	-	1	1	4	86	PHI	3	23	-	-	1	-
38	HRV	6	86	-	-	3	2	87	URY	5	22	-	-	-	1
39	ARG	6	85	-	1	3	-	88	TUN	4	20	-	-	-	1
	CZE	6	85	-	1	1	3	89	HND	2	17	-	-	-	2
	HEL	6	85	-	-	2	4	90	GTM	4	16	-	-	1	-
42	GEO	6	84	-	-	5	1		LIE	2	16	-	-	-	1
43	ESP	6	82	-	-	3	3		VEN	2	16	-	-	-	1
44	SAF	6	79	-	1	-	4	93	PRI	3	9	-	-	-	-
45	COL	6	77	-	2	-	1	94	SAU	6	8	-	-	-	-
46	SVK	6	76	-	-	3	1	95	BOL	5	5	-	-	-	-
	TKM	6	76	-	-	4	1		UAE	4	5	-	-	-	-
48	AZE	6	74	-	-	3	-	97	KWT	5	3	-	-	-	-
	MDA	6	74	-	1	-	2								

50th IMO 2009 - Germany

	Country	P	TS	G	S	B	H		Country	P	TS	G	S	B	H
1	CHN	6	221	6	-	-	-	53	SVK	6	73	-	-	2	3
2	JPN	6	212	5	-	1	-	54	MNG	6	72	-	-	3	1
3	RUS	6	203	5	1	-	-	55	ESP	6	71	-	-	4	-
4	KOR	6	188	3	3	-	-	56	SWE	6	70	-	-	2	4
5	PRK	6	183	3	2	1	-	57	DEN	6	68	-	1	1	1
6	USA	6	182	2	4	-	-	58	BGD	6	67	-	-	2	3
7	THA	6	181	1	5	-	-	59	AUT	6	66	-	-	2	2
8	TUR	6	177	2	4	-	-	60	LUX	6	65	-	-	3	1
9	GER	6	171	1	4	1	-	61	BIH	6	63	-	-	1	3
10	BLR	6	167	1	4	1	-	62	LVA	6	61	-	-	1	3
11	ITA	6	165	2	2	2	-	63	NOR	6	60	-	-	2	2
	TWN	6	165	1	5	-	-	64	ARM	6	59	-	-	2	1
13	ROU	6	163	2	2	2	-	65	SVN	6	58	-	-	1	3
14	UKR	6	162	3	1	2	-	66	NZL	6	53	-	-	1	3
15	IRN	6	161	1	4	1	-	67	FIN	6	49	-	-	-	4
	VNM	6	161	2	2	2	-		MAC	6	49	-	-	1	2
17	BRA	6	160	1	3	2	-	69	CYP	6	45	-	1	-	2
18	CAN	6	158	1	3	2	-	70	CHI	4	41	-	1	-	-
19	BGR	6	157	1	3	2	-	71	EST	6	40	-	-	-	3
	UNK	6	157	1	3	2	-	72	CRI	4	34	-	-	1	1
	HUN	6	157	1	2	3	-	73	KGZ	6	33	-	-	-	3
22	SRB	6	153	1	3	1	-	74	MAR	6	32	-	-	-	-
23	AUS	6	151	2	1	2	1	75	MAS	2	31	-	1	-	-
24	PER	6	144	-	4	2	-	76	TTO	6	28	-	-	-	2
25	GEO	6	140	-	3	2	1	77	TUN	5	27	-	-	1	-
	POL	6	140	-	2	4	-	78	ECU	6	26	-	-	-	1
27	KAZ	6	136	-	3	3	-		ISL	6	26	-	-	-	1
28	IND	6	130	-	3	2	1		PHI	4	26	-	-	1	-
29	HKG	6	122	1	2	2	-	81	ALB	6	24	-	-	-	-
30	SGP	6	116	-	2	3	1		HND	3	24	-	-	1	-
31	FRA	6	112	-	1	3	2	83	MNE	4	23	-	-	-	1
32	HRV	6	110	-	1	4	1		PRI	6	23	-	-	-	1
33	POR	6	99	-	1	3	2	85	CUB	1	21	-	-	1	-
34	TKM	6	97	-	1	3	-		LIE	2	21	-	-	1	-
35	ARG	6	93	-	1	1	2		PAK	5	21	-	-	1	-
36	AZE	6	91	-	1	2	2		URY	6	21	-	-	-	1
	MKD	6	91	-	1	3	1	89	IRL	6	20	-	-	-	-
38	BEL	6	89	-	1	2	1	90	NGA	6	17	-	-	-	-
39	COL	6	88	-	1	2	2	91	GTM	4	14	-	-	-	1
40	CZE	6	87	-	1	2	3		KHM	6	14	-	-	-	-
41	HEL	6	86	-	-	3	3		PAR	4	14	-	-	-	-
42	UZB	6	85	-	1	2	1	94	SLV	3	13	-	-	-	-
43	IDN	6	84	-	-	4	1		VEN	2	13	-	-	-	-
	SAF	6	84	-	-	2	4	96	PAN	1	12	-	-	-	1
45	TJK	6	82	-	1	2	-	97	BOL	3	9	-	-	-	-
46	ISR	6	80	-	-	3	2	98	MRT	6	8	-	-	-	-
47	NLD	6	79	-	1	1	2	99	SYR	5	7	-	-	-	-
	SUI	6	79	-	-	3	2	100	ZWE	2	5	-	-	-	-
49	LTU	6	77	-	1	1	3	101	BEN	2	3	-	-	-	-
50	MEX	6	74	-	-	3	1		KWT	4	3	-	-	-	-
	MDA	6	74	-	-	4	-		UAE	5	3	-	-	-	-
	LKA	6	74	-	-	2	3	104	ALG	4	2	-	-	-	-

Chapter 12
Memories

12.1 Mircea Becheanu

Romania

Mircea Becheanu is Professor of Algebra at the University of Bucharest. He taught lectures in Algebra and Algebraic Geometry until his retirement in 2009. In 1988 and then between 1995 and 2002 he was the leader of the Romanian IMO team and the president of the Romanian Mathematical Olympiad. In 2001 he published a book containing all problems and solutions, and a short history of all IMOs from 1959 until that time. He made an important contribution to the organization of the 40th IMO in Bucharest. As a member of the IMO Advisory Board he contributed to the growth of the IMO and to the increase of its popularity  *in many countries. His aim is to maintain IMO as a competition open to all talented students and all countries.*

It is well known that First International Mathematical Olympiad was organized by Romania in 1959. The event held between 23-31 July, so we celebrate these days the 50th anniversary of the first and largest international scientific contest for high school students. It is a good opportunity for me to remember some facts regarding the conditions in which this competition was born. I also would like to say a few words about the contribution of the IMO to the improvement of mathematics teaching, all these years.

It is well admitted that the person who had the idea to organize an international mathematical competition for high school students was Tiberiu Roman, at the time the General Secretary of the Romanian Mathematical Society (RMS), whose official name that time was Romanian Society for Mathematical and Physical Sciences (SSMR). That is why he was invited to give a speech at the opening ceremony of the 40th IMO organized in Bucharest in 1999. Inside the Society, the idea was strongly supported by the President, the well known mathematician Grigore Moisil, a Member of the Romanian Academy. Grigore Moisil was a great personality and he impressed the political authorities with this idea. Using his charm and his skill he obtained their agreement and support for the event. We have to take in considerations that those years Romania was a country in the communist camp and everything was under the state supervision. The external cooperation between Romanian Mathematical Society and similar Organizations was almost impossible without this control. This explains why the IMO was born as a competition between former European socialist countries. Therefore, the first countries to take part in IMO were Bulgaria, Czechoslovakia, East Germany, Hungary, Poland, Romania and USSR.

An important point nowadays is to explain why SSMR decided to initiate and organize such a competition. The first reason is the fact that the National Mathematical Olympiad organized by SSMR was very popular in Romania. The Mathematical Olympiad replaced after the World War II another internal competition organized for problem solvers since 1897. Many young students were also very active as problem solvers for Gazeta Matematica, a journal of elementary mathematics continuously published since 1895. For certain, SSMR wanted to offer something more to a generation of talented people - the possibility to challenge their skills with similar students from other countries. Also, the opportunity to exchange ideas with mathematicians from other countries has been taken in consideration, as this was a point in the teacher's program.

A key problem for set up the IMO was to create the Regulations in such a way that they are acceptable for all participants. A sketch was send to all invited countries and after receiving suggestions the final document has been prepared by SSMR. Some of its foresights are, more or less, still valid. I like to mention some of them. The IMO is a competition between students, individually and no team ranking is official. The aim of the IMO is to encourage young people to develop their mathematical skill and to stimulate problem solving in participating countries. All participating countries have right to propose problems and the evaluation of the papers has to be done done by the jury. It is important to mention that during early IMOs it appeared the idea of a jury. All activities were done by the International Committee of the IMO, a body containing all leaders. They also decided about the distribution of the medals, but then there was prizes. In the first IMO, 21 students have been awarded, among 52 participants. The rate is approximately as larger as it is today and this shows that it was a good decision. The organizers decided to offer a complementary program for the students including visits in touristic resorts, schools, museums and others.

A few words about the competition. It is a tradition from First IMO to have a two day examination, but during the years the number of proposed problems has been changed. In the first IMO six problems have been proposed and in the second IMO there were seven. Nowadays, the 1959 problems may be considered rather easy problems. But we have to consider the lack of experience and tradition. I have behind my eyes the list of awarded students of this IMO and I can see that not many had a perfect score. This shows that the examination was appropriate. A characteristic of the early IMOs was the dominance of the computational problems. Year by year countries started to propose more and more interesting problems. New topics like Number Theory and Discrete Mathematics appeared and they gave a new flavor to the competition. On the other hand, Solid Geometry which was well represented that time, is actually not at all used in the examination papers. As I remember, the last problem from Solid Geometry appearing in the Shortlist was in 1999 and it was simplified by the jury to Plane Geometry to be given in the examination in that form.

To continue the history, the second IMO was also organized in Romania between 18-25 July 1960. Only five countries came in this IMO. Poland and USSR did not take part. Each country had a team of 8 students, a leader and a deputy leader. Among 40 participants, 19 have been awarded in the following ratio: 4 First Prize, 4 Second Prize, 4 Third Prize and 9 Honorable Mention. The participating countries considered IMO as a very useful activity and decided to organize it permanently and in all countries. So, the third IMO was organized in 1961 in Hungary. The future of the IMO was strengthened.

When we see the present success of the IMO and its popularity all around the world, we have to recognize the efforts and ingenuity of the first organizers. Year after year, IMO increased in popularity and quality. The Regulations of the IMO have been completed and improved with the aim to create the best competition. IMO problems changed the priorities of mathematical education in many countries. A lot of good books appeared in many languages to help students to become good problem solvers. Last years, internet had an important contribution to make the competition more and more popular. I really hope that IMO will remain for many years an outstanding competition open to all young people which like mathematics.

12.2 István Reiman

Hungary

István Reiman was the trainer of the Hungarian IMO team starting from the first IMO in 1959 for more than four decades. He was 25 times the Deputy Leader of the Hungarian IMO team between the years 1964-1992. He is the author of one of the most comprehensive books on the history of the IMO, originally written in Hungarian, but also available in an English translation.

In the spring of 1959 the Mathematical and Physical Society of Romania on the occasion of a jubilee invited from each country of the socialist block eight students in their final high-school year to participate in a mathematics competition in the city called at that time Oraşul Stalin (now Braşov). The eight students were supposed to be accompanied by two teachers, the list of the students and teachers was to be sent to the organizers in advance. We selected our team on the basis of the results of the Hungarian national competitions and their performance in the monthly problem-solving competition of the "Középiskolai Matematikai Lapok" ("High-school Journal of Mathematics"). After having sent the required list we received a phone-call from the organizers claiming that the name of Béla Bollobás was certainly included erroneously as he was only in the 10th form. The reply was that this was not an error, which was amply proved by the results of Bollobás at this and subsequent competitions.

The competition was a success, the students enjoyed it (one of them, Ferenc Mezei, who later became a physicist, member of the Hungarian Academy of Sciences and director of the Hahn-Meitner Institute in Berlin shot an excellent film of it). The participants came to the conclusion that this competition should be continued. As nobody volunteered to organize it in 1960, Romania once more assembled the competition, with only 5 countries participating. Here Hungary expressed her willingness to host the 1961 competition. It was at this 1961 IMO that the members of the Jury (the team leaders) set out certain guidelines: this competition is based on the honesty of the participants; the proposed problems are to be brought by the team leaders; it is from these problems that the Jury chooses the problems of the two competition days; the number of points given for each problem is decided by the Jury before the actual contest takes place. The idea was raised of forming an official international organization, acknowledged by the governments of all of the participating countries — luckily this idea was rejected.

The 4th IMO (in Czechoslovakia) and the 5th IMO (Wrocław, Poland) were organized along similar lines, but the Poles had two important innovations. Until that time only the team leaders checked the papers of their students and they decided the amount of points their students' solutions were worth. In Wrocław each student's paper was checked by the team leader of another team as well — this was the first coordination. Another important decision of the Polish organizers was to invite a team from Yugoslavia; a country whose political judgement at that time was somewhat ambiguous. As it later turned out, this step opened he door for the IMO to become a worldwide competition.

Starting from the 6th (Moscow) IMO, the coordination was carried out by mathematicians provided by the host country. A special feature of IMOs 5-6-7-8 for Hungary was the presence of two students who won a prize in all four of them: László Lovász, later recipient of the Wolf prize and now President of the International Mathematical Union and József Pelikán who has been the Chairman of the IMO Advisory Board for the last 7 years and team leader of the Hungarian IMO team since 1988.

The 9th IMO was organized by Yugoslavia in the old capital of Montenegro, Cetinje. This was the moment of the opening to the world: teams from the United Kingdom, France, Italy and Sweden participated. A new welcome extension for us, Hungarians was the 12th IMO, organized by Hungary: our long-awaited Western neighbour, Austria took part for the first time. For me personally, a consequence of this development was the fact that in 1976 my most memorable IMO took place in Austria, in Eastern Tirol in Lienz: marvellous landscapes, marvellous organization.

In 1982 Hungary was again host of the IMO with only teams of 4 students participating; despite this fact the competition from the mathematical point of view — thanks to the contributing strong team of mathematicians — was a real success. Politics has seldom intervened in IMO affairs; whenever this happened it either meant not inviting some countries, or some countries not sending teams. This happened to us in 1978 when along with the teams of the Soviet Union and East Germany our team was missing from the 20th IMO held in Bucharest. The reason was similar to the one preventing most socialist countries to participate in the 1984 sports Olympics in Los Angeles. Fortunately, in 1982 we managed to arrange the invitation of Israel, despite the fact that we had no diplomatic relations at that time.

An important step forward in the history of the IMOs was the establishment of the Site Committee resp. Advisory Board. This has the task among other things to guarantee the continuity of the IMO. It is a great satisfaction and pleasure to me that the Advisory Board has been chaired in the last 7 years by József Pelikán. The original reason leading to the formation of the Site Committee was the fact that in the year 1980 no IMO was held. In those years the next host country was picked in a somewhat haphazard manner, in some cases it was only a few weeks before the IMO that the actual host country was definitively determined. In 1979 there was hearsay about Mongolia organizing the next IMO (this was published as a fact in the American Mathematical Monthly), finally no invitation came forward. There were a couple of regional competitions organized as a substitute for the IMO. We were invited by Erkki Pekkonen to Finland along with the teams of Great Britain and Sweden to a superbly organized olympic-style competition.

For me personally one of the most important activities and source of joy in my life was the IMO. Starting in 1961 for more than 40 years I organized and led the training and preparation of the Hungarian team for the IMO. During this time I met many outstanding talents from whom I learned a lot. (From among the participants of the first 15 IMOs 19 are at present members of the Hungarian Academy of Sciences.) In the beginning Hungary had the advantage of having the longest experience in mathematics competitions: since 1894 we have a continuous series of nationwide mathematics competitions and since 1896 we have a high-school mathematics journal.

An important pedagogical aspect of the preparation to such competitions is to awake the students to the consciousness of the fact that an eventual failure does in no way diminish his/her mathematical appraisal: many outstanding mathematicians were no good competitors. On the other hand good competition results are signs of talent and hard work. There are many places where competitions are considered as the only way of nurturing mathematical talent, partly because competitions are relatively easy to organize. This prevents many students from developing their talents; it should be clear that there are other fruitful ways to unfold mathematical talent.

The Hungarian team always finished in the first half of the field, often among the very top teams. On this achievement a shadow is cast by the fact that we did not manage to draw attention to our results nationwide: our IMO participants come from about 2–3 % of all our high-schools and even those students who come anywhere close to IMO participation come from about 6–8 % of the high-schools.

From 1964 to 1992 (with the exception of three years) I was the deputy leader of the Hungarian IMO team; I got to know several colleagues, several countries. It gives me a pleasure to see that these days the IMO is a true world event, perhaps the most significant encounter of talented youth from all over the world, and all this on the basis of mathematics! Nobody could dream of such an extension of the IMO in the beginnings. It is with great respect that I remember those people who in the beginnings nurtured and spread the idea of the IMO. Among them I would like to mention in particular Alipi Nikolov Mateev and Stoian Budurov from Bulgaria, Rudolf Zelinka, Ján Vyšin and František Zítek from Czechoslovakia, Mieczysław Czyżykowski and Andrzej Mąkowski from Poland, Endre Hódi from Hungary, Johannes Gronitz, Herbert Titze and Wolfgang Engel from Germany (GDR), Tiberiu Roman and Gheorge D. Simionescu from Romania and Elena Alexandrovna Morozova and Ivan Semionovich Petrakov from the Soviet Union.

I do not consider it my task to evaluate the effect of the IMO on the development of mathematics in the world, but the existence of such an effect is beyond any doubt. It is to be hoped that we will continue to experience such an effect and many more pleasant memories await us in future IMOs.

12.3 Wolfgang Engel

Germany

Wolfgang Engel is Professor of Theoretical Mathematics (Algebra and Geometry) at the University of Rostock. Besides several positions at the university before his retirement in 1993 he is well known for his many activities related to mathematics at secondary and high schools. He was member of the Paedagogical Academy of Sciences of the GDR and is the father of the Mathematik-Olympiaden in the GDR and consequently in Germany. For many years he was the president of the committee for the Mathematik-Olympiaden and later he was the president of the Mathematische Gesellschaft of GDR. At the two IMOs in the GDR (1965 in Berlin and 1974 in Erfurt) he was the Chairman of the International Jury. Under his leadership there were two breakthroughs in IMO history: the first participation of a non-socialist country (Finland) in 1965 and the first participation of the USA in 1974. In 1998 he was honoured with the Paul-Erdös-Award of the World Federation of National Mathematics Competitions.

The first student mathematics competitions were held in the 19th century at the Philantropinum grammar school in Dessau. In 1957, the periodical "Mathematik, Physik und Chemie in der Schule" published articles in the German Democratic Republic (GDR) in which the extracurricular activities and competitions in mathematics in Hungary (the Eötvös Competition, since 1894) and in the Soviet Union (the Leningrad Mathematical Olympiad, since 1934) were reported. Through a lecture and study trip to Hungary in 1958, I had the opportunity to get to know the competitions there.

Around 1960, a worldwide discussion about the modernization of mathematical instruction began that soon led to changes in school curricula. A broad interest in mathematics was thereby awakened. In 1959, the first International Mathematical Olympiad (IMO) took place in Romania under the direction of Prof. Gh. H. Moisil, at which eight students from each of Bulgaria, the GDR, Poland, Romania, the Soviet Union, Czechoslovakia and Hungary participated. Since GDR students were selected only on the basis of their outstanding high school examinations, but weren't in any way prepared, they didn't do very well. Romanian and Hungarian students achieved the best results. The government of the GDR henceforth supported all measures that led to improvements in instruction and extracurricular advancement, including mathematics competitions (the so-called "Mathematikbeschluss" of 1962).

The first *Olympiade Junger Mathematiker der DDR* was organized through the Ministry of National Education in the 1961/62 school year. The name *Olympiade* was chosen based on its use in other countries, although it doesn't have the meaning of the Greek word. In June of 1962, the Mathematical Society of the GDR was founded, which, among other things, set as a goal the support of the undertaking of Mathematical Olympiads. It, together with the Ministry of National Education, was a supporter of the *Olympiade Junger Mathematiker der DDR*. Their goals were:

- to help see to it that students acquire solid knowledge and skills in the field of mathematics inside and outside of the classroom, expand their knowledge base, and are trained to think mathematically,
- to make the growing importance of mathematics in the shaping of society known to all students,
- to awaken and deepen interest or even fascination for the subject of mathematics in the majority of students,
- to identify students interested and gifted in mathematics, so that their systematic advancement can be effected.

They were implemented every school year in four rounds: first as a homework competition in the schools (with a time frame of about one month), then as written examinations in the 218 counties (over one day), in the 15 districts (over two days) and finally for the whole GDR (over two days). The best from one round could take part in the next. From the 1962/63 school year on, the four rounds of each Mathematical Olympiad were organized approximately as they are today. The number of participants was very high. In 1966/67, 987 000 participants were registered in the first round. This was about 75% of all applicable students. The written examinations in 1966/67 had 50 000 participants in the second round, 2770 in the third round and 240 in the fourth round. From the circle of winners of this round, 12 to 14 candidates were selected for a ten-day preparatory course for the IMO, of which eight ultimately formed the GDR team. The Olympiads were met with broad public interest. Newspapers, radio broadcasters, television reporters and newsreels reported in detail about the problems, participants and program of the competitions.

Since my interest in questions of mathematics education as a trained high school teacher was well known, in 1962 the founding board, through its chair, Prof. Dr. Kurt Schröder (rector of the Humboldt-Universität Berlin), and the Minister of National Education, Prof. Dr. A. Lemnitz, asked me to assume the chair of the Central Committee for Mathematical Olympiads, which I held until 1974. I was thereafter, until 1981, connected to the Olympiads as chair of the Mathematical Society and then as commissioner of the board. The secretary (a kind of manager) of the committee was for many years Chief Lecturer Herbert Titze (†) from Berlin, who contributed significantly to the development of the Olympiads and alongside other mathematics teachers oversaw the GDR team at the IMOs in the first four years. Directing the Problem Commission, which developed the problems and sample solutions, was the responsibility of Prof. Dr. Udo Pirl (†, Humboldt-Universität Berlin). He also prepared the IMO problems in 1965 and 1974. After 1990, the Olympiads were continued via various initiatives. Since 1994, the association *Mathematik-Olympiaden e.V.* has been responsible for the competition, in which all German states participate and which is held under the auspices of the Federal President.

But we only ever viewed the Olympiads as a stimulus to inspire students to do mathematics. Extracurricular activities like workshops, clubs, camps and student societies for mathematics therefore played a large role in the GDR. Many mathematics teachers and high school faculty members got involved. For their work, they received overtime pay, small honorariums, bonuses or government awards. The fundamentals covered in the classroom were also improved. In five special classes for mathematics and physics or chemistry at colleges and universities, as well as in 14 special schools focused in mathematics, natural sciences, and engineering, gifted and interested students were given an advanced education, which led to success.

I was the delegation leader of the GDR team for the first time in 1963, at the fifth IMO in Poland, at which eight countries with a total of 64 students participated. The delegation leaders and students assembled in Warsaw. The jury, under the direction of Prof. St. Straszewicz (Warsaw), convened at the university. In those years, it was comprised of Prof. A. N. Matthèev (Sofia), Prof. M. Ilič-Dajovič (Belgrade), Prof. M. Czyzykowski (Warsaw), Prof. T. Roman (Bucharest), Doz. E. A. Morosowa (Moscow), Prof. J. Vyšin (Prague), Prof. E. Hody (Budapest) and me. After two days of discussions, the problems were approved, translated into the languages of the students and copied. The schedule of events took us to the surroundings of Warsaw. Then we took the train to Wroclaw (formerly Breslau). There the students worked on the problems over two days. The delegation leaders then corrected their solutions. The GDR students achieved three third prizes (today bronze medals) and placed second to last in the unofficial country rankings.

In 1964, the IMO took place at Moscow State University on the Lenin Hills, under the direction of Prof. A. J. Markuševič. The Mongolian People's Republic first participated here. With one second prize and two third prizes, the GDR students achieved a somewhat better result.

In 1965, the nationally owned publisher Volk und Wissen began planning the mathematical student magazine "Alpha", whose first issue appeared in 1967 under the editorship of the lecturer J. Lehmann (Leipzig). It became a widely read periodical that inspired many students to do mathematics. In preparation for the first IMO on German soil in 1965, the logo, in commemoration of an accomplishment made by C. F. Gauss in his youth, contained a 17-gon, compass and straight edge. This logo also appeared in 1977 on the GDR stamp commemorating Gauss's 200th birthday. It is still used today in a modified form. At the seventh IMO, Finland represented the first "non-socialist" country to participate. The jury, under my direction, gathered in a Berlin hotel, and the students were accommodated at the Jugendhochschule am Bogensee (on property formerly owned by J. Goebbels), north of Berlin, where the exams were also taken. The schedule of events included sightseeing in Potsdam and Berlin as well as a trip to Naumburg, Weimar, Karl-Marx-Stadt (today Chemnitz) and Dresden. The award ceremony took place in the congress hall at the Alexanderplatz. With two second prizes and three third prizes and fifth place in the unofficial country rankings, a further improvement was achieved. We were especially impressed by the Hungarian participants L. Lovász and J. Pelikán, who between 1963 and 1966 attained top places (each had three first prizes and one second prize).

Between 1965 and 1990, the professors H.-J. Weinert (Potsdam), H. Bausch (Berlin), and G. Burosch (Rostock) acted as delegation leaders of the GDR team. In the years from 1966 to 1975, the GDR team ranked among the first three places in the unofficial country rankings, and in 1968 took first place. W. Burmeister, who between 1967 and 1971 achieved three first prizes and two second prizes, and S. Heinrich, who between 1966 and 1969 achieved two first prizes and two second prizes, were the most successful participants of the GDR team. Worth mentioning is also P. Kröger, who as a seventh grade student at the 14th IMO in 1972 achieved a full score (40) with a first prize and a special prize. It was arranged that several successful Olympiad participants attend university lectures in mathematics, skip midterm exams and work on small research projects with professors. Especially successful IMO participants received special stipends upon beginning their studies.

Although their respective ministries had reservations, several delegation leaders aspired to invite more countries to the IMOs. In 1966, during the International Congress of Mathematicians in Moscow, I had a talk with Prof. N. D. R. Turner (Albany), who represented the American High School Mathematics Examination. In 1969, during a lecture trip to Austria, I spoke with Prof. W. Nöbauer, at that time chair of the Austrian Mathematical Society, about the Olympiads and gave him informational material.

When the 16th IMO took place in 1974, again under my direction in the GDR, 18 countries were already taking part: Bulgaria, the GDR, Finland, France, Great Britain, Yugoslavia, Cuba, Mongolia, the Netherlands, Austria, Poland, Romania, Sweden, the Soviet Union, Czechoslovakia, Hungary, and, for the first time, the USA and Vietnam. At this IMO, we composed the first printed set of regulations, which essentially still hold today. Students, along with their chaperones, were accommodated in a dormitory at the Pädagogische Hochschule Erfurt, and members of the jury were accommodated at the hotel "Elephant" in Weimar. The schedule of events included sightseeing in Erfurt, Eisenach (Wartburg), Weimar, Potsdam and Berlin. The closing ceremony also took place in the congress hall in Berlin. The best team was the Soviet Union with 256 out of 320 points. With 236 points, our students reached fourth place and achieved five second prizes and two third prizes.

At this IMO, the following incident occurred: Upon the arrival of the teams at the campus of the Pädagogische Hochschule, their respective national flags were raised. During the night following the arrival of the American team, their flag was stolen. Since there was no adequate replacement in Erfurt and its surroundings, a car was sent to Leipzig, where, because of the exhibitions taking place there, a suitable American flag was available. Several hours later, while the driver was returning, the police found the culprit with the flag (a drunk). By morning, all flags could again be flown. None of the guests had noticed a thing.

Nothing is known about what became of the participants of the first IMO in 1959. Of the 91 students from the GDR who participated in the IMO from 1960 to 1976 (not a few students were participants at several Olympiads; four became ill and three died before completing a degree), at least 38 habilitated, and of these 28 became professors. At least 22 further students earned a doctorate in mathematics, physics or engineering, and two in medicine or veterinary science. One became president of a German state, one an actor (a cabaret artist), one a musician at the Hessische Staatstheater Wiesbaden and one made television appearances as a human calculator.

In the "old" Federal Republic of Germany, the Stifterverband für die Deutsche Wissenschaft first organized the *Bundeswettbewerb Mathematik* (in three rounds) in 1970. In the first two rounds, students solved problems at home over the course of two months. The winners of the first and second prizes of the first round received the problems for the second round. The first prize winners of this round were invited to a colloquium with mathematicians, at which the national winners were determined. A substantial push for this was made by Prof. A. Engel (no family relation to me!), who in 1965 already reported about competitions in the periodical "Der Mathematische und Naturwissenschaftliche Unterricht" and who published the book "Olympiadeprobleme aus der UdSSR". There was an exchange of letters and problems between him and the chair of the board of trustees, as well as a meeting with the then doctoral candidate H. Sewerin. In 1977, a team from the Federal Republic of Germany under the direction of Prof. A. Engel first participated at the 19th IMO in Belgrade. The 30th IMO in Braunschweig took place under his presidency.

In 1982, 30 countries already participated in the 23rd IMO. In order to better handle organizational problems, the number of students each country was allowed to send was reduced. This number is now six. At the 49th IMO in 2008, 97 national teams sent a total of 535 students.

Under my chairmanship, the Mathematical Society of the GDR, together with the Ministry for Universities and Professional Schools, organized a competition called "Wissenschaftliche Studentenkonferenz" in 1977, at which students were to present their first research results. Many former Olympiad participants took part in this. The best of this competition received awards. Today, these student conferences are continued by the Deutsche Mathematiker-Vereinigung.

12.4 Radu Gologan

Romania

Radu Gologan is a professor at the Politechnical University in Bucharest and a researcher at the Institute of Mathematics of the Romanian Academy. His field of interest in research is ergodic theory and theoretical statistical physics. In 1970-71 he was a member of the Romanian team for the IMO obtaining a bronze and a silver medal. Since 2001 he is the coordinator of the Romanian Mathematical Olympiad and the leader of the Romanian Team at the IMO's. At present he is the president of the Romanian Mathematical Society.

Welcome address at the 50th IMO anniversary celebration by the representative of Romania – the country, which initiated the IMOs

Damen und Herren, Kollegen und Studenten! Dear colleagues, dear students!

The idea of an international competition in mathematics for the young belongs to the Romanian Mathematical Society, which will be 100 years old next year. Also, the first two IMO's took part in Romania. These facts granted me the great honor to address this assembly, which marks the 50th anniversary of the creation of the IMO. And, though the IMO is older than most of the people present here, it is young in spirit and with a bright future.

A lot of nice or important things can be said about this extraordinary event, which influenced not only the way mathematics is taught around the world, but also contributed to the increase of the general interest in this domain, illustrated, among other, in the augmentation of the number of math researchers. In the sequel, I will briefly point three facts, which I consider important.

1. We all have to pay homage to a Romanian professor, Tiberiu Roman, who imposed the idea and had the energy to organize the first IMO, using his connections and relying mostly on his work.
2. During the "Iron Curtain" era, the IMO was one of the few possibilities to put together clever minds from different countries, in a period when contacts where hardly possible.
3. Friendly relations which started when young people were participants in the IMO still continue today, both at the personal and at professional level.

Nowadays, the IMO becomes more and more popular. In order to maintain this tendency, some important moves should be done, in sustaining the newcomers in the preparation of their participation in the competition. The IMO Advisary Board must try to access international organizations which are willing or able to support this activity.

In the end, I will submit you two quizzes.

1. Do you know why the title of the competition is "Olympiad" instead of the correct word "Olympics"??
2. Which countries hosted the "zero ending", i.e. the 10th, the 20th, the 30th, the 40th and the 50th edition of the contest? And which country should organize the 60th edition?

And now, the answers:

1. A Romanian translator used the improper form in the first official title.
2. Romania and Germany, and I will let you guess the last answer.

Chapter 13
Hall of Fame

Many IMO participants have, in their later careers as research mathematicians, obtained some of the most prestigious awards for mathematical research – and many successful research mathematicians had, in their youths, participated at IMOs. We list several of the most renowned awards in mathematics, together with those recipients that participated at IMOs.

We also list those individuals that received, during their youths, at least three gold medals at IMOs.

Fields Medals

The Fields Medal is often viewed, in terms of reputation, as the greatest honor a mathematician can receive. Every four years at the International Congress of Mathematicians, up to four Fields medals are awarded to "young" mathematicians (at most 40 years old). The first Fields medals were awarded in 1936. It comes with a gold medal and monetary award (approx. € 10 000 in 2006).

Year	Name	Country	IMO	Medals
1978	Margulis, Grigorij A.	USS	1962	S
1990	Drinfel'd, Vladimir	USS	1969	G
1994	Lions, Pierre-Louis	FRA	1973	
1994	Yoccoz, Jean-Christophe	FRA	1973, 1974	G, S
1998	Borcherds, Richard E.	UNK	1977, 1978	G, S
1998	Gowers, W. Timothy	UNK	1981	G
2002	Lafforgue, Laurent	FRA	1984, 1985	S, S
2006	Perelman, Grigori	USS	1982	G
2006	Tao, Terence	AUS	1986, 1987, 1988	G, S, B
2010	Lindenstrauss, Elon	ISR	1988	B
2010	Ngô, Bao-Châu	VNM	1988, 1989	G, G
2010	Smirnov, Stanislav	USS	1986, 1987	G, G

Nevanlinna Prizes

The Nevanlinna Prize honors great achievements in mathematical aspects of computer science; it was first awarded in 1982. Like the Fields Medals, the Nevanlinna Prizes are awarded every four years, at the International Congress of Mathematicians. It is awarded to a single recipient of age at most 40 years. It comes with a gold medal and a monetary award similar to that of a Fields medal.

Year	Name	Country	IMO	Medals
1990	Razborov, Alexander	USS	1979	G
1998	Shor, Peter W.	USA	1977	S

Wolf Prizes for Mathematics

Wolf Prizes for Mathematics have been awarded annually since 1978 to two mathematicians. It comes with a diploma and a monetary award of $100 000. At least until the creation of the Abel Prize in 1999, the Wolf Prize was viewed as a mathematics analog to the Nobel prize (because the Fields medal, while more prestigious, is awarded only every four years and only to mathematicians under 40).

Year	Name	Country	IMO	Medals
1999	Lovász, László	HUN	1963–1966	G, G, G, S
2005	Margulis, Grigorij	USS	1962	S

Gödel Prizes

The Gödel Prize is a prize for outstanding papers in theoretical computer science, named after Kurt Gödel and awarded jointly by the European Association for Theoretical Computer Science (EATCS) and the Association for Computing Machinery Special Interest Group on Algorithms and Computation Theory (ACM SIGACT). The Gödel Prize has been awarded annually since 1993. It includes a monetary award of $5 000.

Year	Name	Country	IMO	Medals
1993	Babai, László	HUN	1966–1968	G, S, S
1994	Håstad, Johan	SWE	1976–1977	G, B
1999	Shor, Peter	USA	1977	S
2001	Lovász, László	HUN	1963–1966	G, G, G, S
2007	Razborov, Alexander	USS	1979	G

Salem Prizes

The Salem Prize is awarded annually to a young mathematician who has done outstanding work in the field of interest of Raphael Salem, primarily the theory of Fourier series (or, more broadly, analysis).

Year	Name	Country	IMO	Medals
1978	Björn E. Dahlberg	SWE	1967	
1982	Alexei B. Aleksandrov	USS	1970–1971	S, S
1988	Jean-Christophe Yoccoz	FRA	1973–1974	G, S
1990	Sergei Konyagin	USS	1972–1973	G, G
1999	Fedor Nazarov	USS	1984	G
2000	Terence Tao	AUS	1986–1988	G, S, B
2001	Stanislav Smirnov	USS	1986–1987	G, G
2003	Elon Lindenstrauss	ISR	1988	B
2003	Kannan Soundararajan	IND	1991	S
2005	Ben Joseph Green	UNK	1994–1995	S, S
2006	Artur Avila de Melo	BRA	1995	G
2007	Akshay Venkatesh	AUS	1994	B
2008	Boçz Klartag	ISR	1996	S

Clay Research Awards

The Clay Mathematics Institute presents the Clay Research Award annually (since 1999) to recognize major breakthroughs in mathematical research. The award usually goes to two mathematicians, or to two groups of mathematicians. Awardees receive the bronze sculpture "Figure Eight Knot Complement VII/CMI" by sculptor Helaman Ferguson.

Year	Name	Country	IMO	Medals
2000	Lafforgue, Laurent	FRA	1984, 1985	S, S
2001	Smirnov, Stanislav	USS	1986, 1987	G, G
2003	Tao, Terence	AUS	1986, 1987, 1988	G, S, B
2004	Ben Joseph Green	UNK	1994, 1995	S, S
2004	Ngô, Bao-Châu	VNM	1988, 1989	G, G

World Federation of National Mathematics Competitions

The David Hilbert Award (until 1996) and the Paul Erdős Award, given by the World Federation of National Mathematics Competitions, honor mathematicians who are prominent on an international or national scale in mathematical enrichment activites. The recipients of these awards are listed below. Most of the recipients were or are closely related to the IMO as team leader, deputy, jury member or former participant.

David Hilbert Award

Name	Country	Year	Name	Country	Year
Barbeau, Edward	CAN	1991	Liu, Andy	CAN	1996
Engel, Arthur	GER	1991	Losada, Maria de	COL	1994
Gardner, Martin	USA	1992	O'Halloran, Peter	AUS	1994
Klamkin, Murray	CAN	1992	Pollard, Graham	AUS	1991
Kuczma, Marcin	POL	1992			

Paul Erdős Award

Name	Country	Year	Name	Country	Year
Andzans, Agnis	LVA	1998	Marinkovic, Bogoljub	YUG	2002
Atkins, Warren	AUS	2004	Mientka, Walter	USA	1994
Bellot Rosado, Francesco	ESP	2000	Reiman, Istvan	HUN	2000
Berzsenyi, George	USA	1996	Reiter, Harold Braun	USA	2002
Chua, Simon	PHI	2006	Rejali, Ali	IRN	2006
Davidson, Luis	CUB	1992	Sanjmyatav, Urgengt.	MNG	1994
Deledicq, André	FRA	2004	Saul, Mark	USA	1998
Dunkley, Ron	CAN	1992	Shian, Leou	TWN	2008
Engel, Wolfgang	GER	1998	Soifer, Alexander	USA	2006
Frauring, Patricia	ARG	2004	Sun, Wen-Hsien	TWN	2002
Gardiner, Tony	UNK	1996	Surányi, János	HUN	2000
Gronau, Hans-Dietrich	GER	2008	Tabov, Jordan	BGR	1994
Henry, Bruce	AUS	2008	Taylor, Peter	AUS	1994
Holton, Derek	NZL	1996	Webb, John	SFA	1992
Konstantinov, Nikolay	RUS	1992	Zonghu, Qiu	CHN	1994

Winners of Three or More Gold Medals

Name	Country	IMO participation	Medals	
Reiher, Christian	GER	1999–2003	GGGGB	
Barton, Reid	USA	1998–2001	GGGG	
Burmeister, Wolfgang	GDR	1967–1971	GGGSS	2 special prizes
Boreico, Iurie	MDA	2003–2007	GGGSS	1 special prize
Härterich, Martin	GER	1985–1989	GGGSB	
Barzov, Vladimir	BGR	1999–2002	GGGS	
Lovász, László	HUN	1963–1966	GGGS	2 special prizes
Nagao, Kentaro	JPN	1997–2000	GGGS	
Nikolov, Nikolay	BGR	1992–1995	GGGS	1 special prize
Pelikán, József	HUN	1963–1966	GGGS	2 special prizes
Scholze, Peter	GER	2004–2007	GGGS	
Soejima, Makoto	JPN	2005, 2007–2009	GGGB	
Badzyan, Andrey	RUS	2002–2004	GGG	
Banica, Theodor	ROU	1989–1991	GGG	
Dourov, Nikolai	RUS	1996–1998	GGG	
Dremov, Vladimir	RUS	1998–2000	GGG	
Goldberg, Oleg	RUS, USA	2002–2004	GGG	
Hornet, Stefan L.	ROU	1997–1999	GGG	
Ivanov, Ivan	BGR	1996–1998	GGG	
Ivanov, Sergej	USS	1987–1989	GGG	
Kralev, Rosen	BGR	2003–2005	GGG	
Liu, Tiankai	USA	2001–2002, 2004	GGG	
Malinnikova, Evgenija	USS	1989–1991	GGG	
Manea, Mihai	ROU	1999–2001	GGG	
Manolescu, Ciprian	ROU	1995–1997	GGG	3× full score
Mazur, Przemyslaw	POL	2006–2008	GGG	
Norine, Serguei	RUS	1994–1996	GGG	
Norton, Simon	UNK	1967–1969	GGG	2 special prizes
Rász, Béla András	HUN	2002–2004	GGG	
Rickard, John R.	UNK	1975–1977	GGG	2 special prizes
Sannikov, Yulij	UKR	1994–1996	GGG	
Terpai, Tamas	HUN	1997–1999	GGG	

Chapter 14
The Golden Microphone

Rafael Sánchez Lamoneda

Rafael Sánchez was the team leader of Venezuela for many years.

Announcing the winner of the Golden Microphone 2008 in Madrid.

It was during the 39th International Mathematical Olympiad in Taipei, Taiwan, when it occurred to me to count down how many times some team leaders speak during the Jury Meetings. It happened because it was obvious that some of us used to speak many times, and some times even after the point was completely clear for everybody present.

After that IMO, I continued the counting year after year, and at the end of the Olympiad, all the Spanish speaking community, and the Portuguese speaking community also, met together after the award ceremony and farewell dinner, and in there I announced the winner of that year. I have to mention that also Claude Deschamps, the French team leader was a regular guest invited to our parties and he announced the winner in French in our private ceremony.

In the year 2002 in Glasgow we took a step ahead, and we invited to our party the winner of that year. I must say I was a little worried about that, but the winner understood the joke quickly, of course, he was an IMO team leader, and he accepted the prize and delivered a short speech.

I do not know how Paul Vaderlind knew about this, perhaps the 2002 winner told him, it is one of the mysteries of this story, but next year in Japan, he approached to me at the beginning of the IMO, and he asked me if it was true I was counting the leaders interventions during the jury meetings. I said yes and he congratulated me because he considered it a very good idea as it contributed to the friendship and good relations between all of us. I must say, in honor to the truth, that until that moment I had not thought about that in such a way, but after Paul told that to me, I realized he was right, so I decided to do the counting more seriously.

Nevertheless, once I was involved in the IMO day by day affairs, I forgot about how seriously I should do the counting, but I did it as usual anyway. The final day, previous to the official closing ceremony, Paul talked to me again and then I decided to do something that really created a good impression in the leaders and the deputy leaders community. First of all I went with María Gaspar, Patricia Fauring, Flora Gutierrez and Joana Teles, to buy a microphone, because until then, the prize was made out of a piece of paper, golden just in case we could find it. We did not get a golden microphone, instead we bought a pink microphone but with some golden reflects. It was what we were able to find in a store close to the auditorium in which the ceremony was supposed to start half an our later.

After the formal award and closing ceremony we had an excellent banquet and again I had forgotten about the prize, in my mind was the idea of having again our private party to give the prize to that year winner, but the persistent Paul was again around and then he came to my table to ask me about the prize, who the winner was and when the award ceremony was supposed to be. An idea came up to my mind. Why not do a very "formal" award ceremony? It was the right time to do it, everybody was happy, and having an excellent time, so I asked Paul to be the Master of Ceremony, and of course he accepted immediately, but it was not the end, I asked Hans-Dietrich Gronau, Nazar Agakhanov, Claude Deschamps, and María Gaspar, the leaders from Germany, Russia, France and Spain, respectively, to do the translations to the official languages, and *voilà*, we had all the ingredients for our "formal" award ceremony.

Since then every year we had our ceremony, even in Mexico, when we also had a hurricane in Merida the final days of that IMO.

With the years of experience we developed unwritten rules to decide who is the winner of the golden microphone. They are very simple indeed. In order to win you have to speak a lot, you have speak as many times as possible. A second rule says

Quintijn Puite wins the Golden Microphone at the IMO 2008 in Madrid/Spain.

the same person cannot be winner in two consecutive years. In case of two or more leaders with the same number of points, the winner will be the one with the longest speech. Of course there always be some subjectivity.

Goeffrey Smith is the winner of the 2006 Golden Microphone.

I would like to say that for me it has been a beautiful experience of friendship. I thank Paul for his support and ideas, and I thank all the team leaders during these years for giving me their understanding.

The Golden Microphone Winners

Taipei, Taiwan	1998	David Hunt	Australia
Bucarest, Romania	1999	David Hunt	Australia
Taejon, Korea	2000	David Hunt	Australia
Washington DC, USA	2001	David Hunt	Australia
Glasgow, Scotland	2002	Michael Albert	New Zealand
Tokyo, Japan	2003	Nicolau Saldanha	Brasil
Athens, Greece	2004	Arash Rastegar	Iran
Merida, Mexico	2005	Arkadii Slinko	New Zealand
Ljubljana, Slovenia	2006	Geoffrey Smith	United Kingdom
Hanoi, Vietnam	2007	Radu Gologan	Romania
Madrid, Spain	2008	Quintijn Puite	The Netherlands
Bremen, Germany	2009	Geoffrey Smith	United Kingdom

Appendix A
IMO Country Codes

Throughout this volume we follow the usual IMO custom of possibly abbreviating official country names by a 3-letter code. Lists and tables are sorted alphabetically, whereby for some country names, an apparently more familiar short form is used.

ALB	Albania
ALG	Algeria
ARG	Argentina
ARM	Armenia
AUS	Australia
AUT	Austria
AZE	Azerbaijan
BAH	Bahrain
BGD	Bangladesh
BLR	Belarus
BEL	Belgium
BEN	Benin
BOL	Bolivia
BIH	Bosnia and Herzegovina
BRA	Brazil
BRU	Brunei
BGR	Bulgaria
KHM	Cambodia
CAN	Canada
CHI	Chile
CHN	People's Republic of China – sorted as: China, People's Republic of
COL	Colombia
CIS	Commonwealth of Independent States
CRI	Costa Rica
HRV	Croatia
CUB	Cuba
CYP	Cyprus

CZE	Czech Republic
CZS	Czechoslovakia
DEN	Denmark
ECU	Ecuador
EST	Estonia
FIN	Finland
FRA	France
GEO	Georgia
GDR	German Democratic Republic
GER	Germany
HEL	Greece
GTM	Guatemala
HND	Honduras
HKG	Hong Kong
HUN	Hungary
ISL	Iceland
IND	India
IDN	Indonesia
IRN	Islamic Republic of Iran – sorted as: Iran, Islamic Republic of
IRL	Ireland
ISR	Israel
ITA	Italy
JPN	Japan
KAZ	Kazakhstan
PRK	Democratic People's Republic of Korea – sorted as: Korea, Democratic People's Republic of
KOR	Republic of Korea – sorted as: Korea, Republic of
KWT	Kuwait
KGZ	Kyrgyzstan
LVA	Latvia
LIE	Liechtenstein
LTU	Lithuania
LUX	Luxembourg
MAC	Macau
MKD	The former Yugoslav Republic of Macedonia – sorted as: Macedonia, The former Yugoslav Republic of
MAS	Malaysia
MRT	Mauritania
MEX	Mexico
MDA	Republic of Moldova – sorted as: Moldova, Republic of
MNG	Mongolia
MNE	Montenegro
MAR	Morocco
MOZ	Mozambique
NLD	Netherlands

NZL	New Zealand
NIC	Nicaragua
NGA	Nigeria
NOR	Norway
PAK	Pakistan
PAN	Panama
PAR	Paraguay
PER	Peru
PHI	Philippines
POL	Poland
POR	Portugal
PRI	Puerto Rico
ROU	Romania
RUS	Russian Federation
SLV	El Salvador – sorted as: Salvador
SAU	Saudi Arabia
SRB	Serbia
SCG	Serbia and Montenegro
SGP	Singapore
SVK	Slovakia
SVN	Slovenia
SAF	South Africa
ESP	Spain
LKA	Sri Lanka
SWE	Sweden
SUI	Switzerland
SYR	Syria
TWN	Taiwan
TJK	Tajikistan
THA	Thailand
TTO	Trinidad and Tobago
TUN	Tunisia
TUR	Turkey
NCY	Turkish Republic of Northern Cyprus
TKM	Turkmenistan
UKR	Ukraine
UAE	United Arab Emirates
UNK	United Kingdom
USA	United States of America
URY	Uruguay
USS	Union of Soviet Socialist Republics – sorted as: USSR
UZB	Uzbekistan
VEN	Venezuela
VNM	Vietnam
YUG	Yugoslavia
ZWE	Zimbabwe

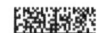